DISCARD

ECOLOGY
A WRITER'S HANDBOOK

Mary Travis Arny, Department of Biology, Montclair State College

Christopher R. Reaske, Department of English, The University of Michigan

ECOLOGY

A Writer's Handbook

With a Full Glossary of Ecological Terms

Random House
New York

Photograph Credits: Page 8—Courtesy of The American Museum of Natural History; Pages 17, 26, 43—George W. Gardner; Page 65—Karin Gurski Batten; Pages 85, 102—Dagmar.

Copyright © 1972 by Random House, Inc.
All rights reserved under International and Pan-American Copyright Conventions. Published in the United States by Random House, Inc., New York, and simultaneously in Canada by Random House of Canada Limited, Toronto.

Library of Congress Cataloging in Publication Data

Arny, Mary (Travis)
   Ecology: a writer's handbook.

   Bibliography: p.
   1. Ecology—Authorship. 2. Ecology—Terminology. 3. Technical writing. I. Reaske, Christopher Russell, joint author. II. Title.
QH541.A72   808.06'6'574502   72-178686
ISBN 0-394-31622-3

Manufactured in the United States of America. Composed by Cherry Hill Composition, Pennsauken, N.J. Printed and bound by The Book Press, Brattleboro, Vt.

Typography by Karin Gurski Batten

First Edition

987654321

ACKNOWLEDGMENTS

We would like to express our thanks to Jim Smith and June Fischbein for coming out to Shelter Island and talking with us about our book. From Random House we also received great help with the manuscript from Zivile Rawson and support from Jess Stein; our thanks to the designer, Karin Batten, should also be recorded. We would like to acknowledge our indebtedness to Nancy P. Arny, of the University of Massachusetts, for meticulously reviewing and revising our glossary of ecological terms. We would also like to thank Mrs. Charles Hemmersley for her help in preparing the manuscript. Finally, we must express our gratitude to Robert and Mary Katharine for being patient with us in general.

M.T.A.
C.R.R.

To Suzanne Travis Reaske and
Katharine Russell Reaske
whose father and grandmother,
among many others, are trying hard
to leave them a better earth.

PREFACE

This book is designed to aid the college student writing about ecology.

We have tried to make the book about good writing in general, as well as about some of the special considerations in writing about ecology. We have included a practical guide to usage and mechanics, and a full glossary of ecological terms. We have purposely made this book brief and practical. By combining the knowledge of an ecologist and biology teacher with that of an English teacher, we have tried to put the main issues, problems, and terminology of ecology into a perspective which will not only help the college student writing about ecology but will also serve the layman seeking to communicate his concern to governmental agencies and officials as well as to the press, in the form of individual and organizational letters.

We have included a glossary of ecological terms to help the writer to be precise, as well as a bibliography that he may turn to as a resource for authoritative facts and materials.

M.T.A.
C.R.R.

CONTENTS

Writing for Your Life — 3

Selecting the Topic — 11

Restricting and Developing the Topic — 19

Technique, Style, and Persuasion — 31

A Glossary of Ecological Terms — 51

Essential Usage — 79

Mechanics — 91

Bibliography — 110

ECOLOGY
A WRITER'S
HANDBOOK

1
WRITING FOR YOUR LIFE

The Ecological Crisis

One car driven one block consumes enough oxygen to keep one hundred people alive one month.

A mature tree will supply the oxygen needed for an adult to live, yet each Sunday edition of *The New York Times* consumes 150 acres of timber.

Pesticide residue is found in every warm-blooded animal. Man could not pass the government's standards; he is not fit to eat, and mother's milk is often not fit to drink. If it were marketed like cow's milk, the F.D.A. would confiscate it.

Downwind from atom testing, infant mortality has risen 50 percent.

There are on earth 3½ billion people and about 3½ billion acres of arable land, and yet 10,000 people die of starvation and malnutrition every single day. Thirty years from now, at the current population increase, there will be about ⅓ acre per person.

The spread of cities takes at least 1½ million acres each year.

Nebraska's underground water supply will be completely dry in 50 to 100 years.

Each New Yorker inhales toxic materials equal to 38 cigarettes a day.

The figures keep coming and coming and coming. Seventy-eight species of mammal, bird, reptile, and fish face imminent extinction, and it is seriously proposed by many ecologists that man should be added to this list of endangered species. In any event, man is clearly faced with an ecological crisis.

The word *ecology* derives from a Greek word meaning household. The study of ecology is the study of our household, the planet Earth, including the entire biosphere. Our planet, in the words of Adlai Stevenson, may be thought of as and is, in actuality, a space ship. All of our resources are on it. In spite of space travel and space fantasy, there is in the foreseeable future no other practical biome or living place for man and the other organisms that inhabit it. We cannot depend upon technology to reverse all of the disastrous effects technology itself has brought us, nor can we hope to create an artificial biome adequate to support even a fraction of the plant and animal population of the globe.

Technology has indeed provided every American with the equivalent in work power of 500 slaves. These slaves are machines. But it must be recognized that slaves, living or mechanical, require fuel in either the conventional terms or food to create energy, that they require oxygen to convert this fuel to energy, and that they excrete wastes.

It is the excrement that has led to the ecological crisis. The "effluent society" has replaced Galbraith's affluent society. Of all of those culpable, the United States of America is the most responsible. It consumes

50 percent of all the raw materials used upon the earth each year in support of only 6 percent of the world's population.

Every human being produces garbage, but it takes one thousand Asians to equal the garbage output of one American. It is for this reason that Americans should not think in terms of how many children they can *afford*, for it is self-evident that the more affluent, the more effluent.

The ecological crisis is brought about by a terrifying population explosion. We simply do not have enough arable land to supply the basic food needs of the world. By the late 1970s we shall be faced with famine beyond our wildest nightmares, and the pacifying prognostication that we can farm the sea or feed the world with grain from improved strains is completely without foundation, and has recently been denied by the very experts who developed these foods.

The Writer's Role

Somebody has to do something about the ecological crisis, and it is beginning to look as though it will have to be the writer who will teach a responsible government and a responsible citizenry to respond without further delay to the need for active and effective control of what is clearly a disastrous situation.

This concerned writer (and teacher) faces a problem he is often unaware of. He assumes that because the facts are obvious to him they are equally obvious to his audience. Few things are more difficult to understand than why somebody else cannot grasp that which is clear to you. An example of this is the simultaneous discovery by two well-known scientists of the earth-shaking fact that overpopulation is the cause of much human misery and specifically of wars. The reader's immediate reaction might be, "Any fool knows that," but apparently any fool does not know it!

A basic cause of our ecological problem lies not only in too large a population but in the fact that a large proportion of that population is driven by a lust for money and convenience and is also too short-sighted to realize that we cannot continue to rape the earth, poison it with offal, and expect it to survive. If the earth dies, we die with it, and the uninformed seem unable to realize that total ecological destruction, the ruin of our air, our soil, our water, the total violation of nature's laws will result in a completely impersonal reprisal that will kill both the just and the unjust with equal indifference. Natural catastrophes are not respecters of persons. After all, Pompeii was a great place to live in while it lasted.

The ecological writer can supply the person who is concerned about dubious practices with facts, facts which are relevant to his immediate environment. Such facts have been and should be presented to the public even if only on a ditto sheet. A word of warning: check your facts thoroughly and be prepared to defend them. The ecological writer who attempts to persuade people to clean up a town will make enemies, and one provable error in his presentation can undermine the entire cause. Often you can get help in the compilation of accurate evidence from garden clubs, service clubs, The League of Women Voters, school or college conservation groups, sportsman's clubs, and even interested individuals.

The average citizen refuses to believe that he is drinking processed sewage, but an astonishing proportion of the population of the United States is doing just that. This same citizen labors under the illusion that primary treatment of sewage such as simple chlorination is all that is required to make it sanitary. Its dilution in the streams dilutes its importance in his eyes. He has no hesitation in drinking from his faucet, but he would not drink the contents of his water closet before flushing—no matter how much chlorine had been added to it. This citizen is certain that the government would not permit his town to provide him with water that is unfit to drink. Little does he know of government! It is the writer's task to make such an individual realize that governments too are susceptible to greed, graft, and political pressure (though the affair of the S.S.T. should have made this obvious).

In writing about ecology you are not writing for fame or for money. You are writing for your life. Converts must be made quickly in order to reverse the disastrous course which we are pursuing, and to create and enforce legislation that will recognize and support the laws of nature. Man as a species is almost environmentally bankrupt. Across the continent from smogged New York lies Los Angeles, and there is plenty of foul air in between. In the city of Los Angeles notices on school bulletin boards warn children: "Do not exercise strenuously or breathe deeply during smog conditions." The peregrine falcon is close to extinct. Collapsed eggs of wild ducks and gulls indicate that they too are beginning to show the effects of DDT, which interferes with the calcium metabolism of birds so that the weight of the incubating female shatters her eggs. What a spiritual loss this constitutes! But the prodigals refuse to face the facts. The ignorant and the careless, the unthinking and the greedy continue to assume that the essentials of life are inexhaustible and that their consumption and destruction for the sake of personal profit and comfort can continue indefinitely.

The supply of life-giving materials is limited. Like any other family faced with economic problems, the family of man will have to choose between tightening its belt and certain disaster. But man is an animal, and a strange one at that. He refuses to recognize danger unless it is

clear and present, which is probably a good thing up to a point. If each of us worried about every eventuality, we would soon become a race of neurotics.

The problem, then, is this: How can we *persuade* our family that the danger of extinction *is* clear and present? We seem to be so constituted that the death of hundreds of thousands of human beings in the Stalinist regime, of millions under Hitler, of more hundreds of thousands in Hiroshima and Nagasaki does not affect us. Perhaps they are too distant, as are the two hundred thousand wiped off the earth by natural catastrophic tides and winds in Pakistan. Our task is to bring the danger into focus and to underline the imminence of catastrophe.

The Purpose of this Book

We believe that it is important for all citizens to write about ecological insanity. Your writing letters is as much a part of ecotactics as our writing this book. The purpose of this book, therefore, is to teach good writing in general, and to teach some of the special considerations in writing about ecology. We seek to help the beginning writer find a topic of interest and importance not only to him but to others, to develop that topic into manageable presentation, and to make that presentation as forcefully persuasive as possible. We discuss ways of finding a topic, ways of restricting and developing a topic, ways of gathering supporting evidence (of crucial importance when writing on ecological problems), ways of applying basic techniques of composition to ecological writing, ways of thinking about such basic matters as style, tone, and audience. While we tend to think a good deal about the writing of essays, we also present advice about other forms of ecological writing such as letters and protest fliers.

In this opening chapter we have sought to introduce to you the extent of the ecological crisis and to argue that it is the writer who finally will be able to do the most about it. The rest of the book, although concerned primarily with the craft of writing, will always reflect that philosophy.

2 SELECTING THE TOPIC

Start Close To Home

Books that deal with the writing of fiction, poetry—in fact, most subjects—advise the writer to stick to topics he knows something about. Ecology is no exception. Leave the pollution of outer space, the smog in Los Angeles, the stinking flow of the mighty Mississippi, the effects of a distant bay of a deep water port to the people who live near them. There are plenty of problems (and thus topics) at home, and besides, ecotactics needs a grass roots army in the field.

It is sad but nevertheless true that Pumpkin Center is not particularly concerned with pollution in Green Patch Village unless and until that pollution shows signs of making trouble in Pumpkin Center. The communications media have only recently begun to grasp the concept that all poor conservation practices, all mismanagement of pesticides, and all pollution are inextricably linked, and that if one village comes apart at the ecological seams the next is sure to follow.

One of the most desperately needed actions is the pinpointing, tabulation, verification, and exposure of pollution at the small town and small stream level. The small stream flows into the brook and the brook into the river. Pollution at the source pollutes everything below it. Thus it makes good sense to select a topic that is close to home for two reasons: first, your proximity to the problem should indicate that you know more about it and can talk about it more authoritatively than a stranger; and second, you will be bringing attention to a problem that perhaps nobody else will have noted.

What kind of problem? Are there, for example, individuals in your town who are dumping their sewage into the local streams without treatment, or are there even small factories releasing chemical sludge or other undesirable effluents into your local streams? Are cattle wading in the brook which runs through the pasture, or are they properly confined to drinking at the farm pond or from troughs? Is the population of gulls on a fresh water pond large enough to be a source of pollution?

Data on local abuses such as these are needed for the completion of the total ecological picture. What is the law in your town about the placement of cesspools relative to wells? Is the distance of cesspools from wells and from streams adequate in view of the type of soil? Are the laws enforced? Do people realize that ground water under a village usually lies there in the form of an underground reservoir or aquifer, and that once polluted this aquifer fouls every well it supplies? These are matters that should be clearly and authoritatively stated, and stated in places easily accessible to the public (including the local press).

In selecting a topic to write about, bear in mind that appeals to government and industry on the exclusive basis of morality or reason often are fruitless. With officials and industrialists, as with most individuals, money talks. Pollution is seldom cleaned up voluntarily. It must be so dramatically demonstrated that the public will place economic pressure on the perpetrator. A case in point is that of a small town in Massachusetts which was dumping raw sewage into a nearby stream. A plot was hatched for Earth Day involving outraged citizens, including many students, who agreed to stand at the outfall and form a bucket brigade that would collect and dump the sewage at the Town Hall. News of the plot leaked, and the town immediately started proceedings to build a proper modern treatment plant. Though this plan involved physical action rather than writing, it is clear that a dramatic attack on such a problem could have utilized a good writer—perhaps to write documented descriptions of the pollution or spirited protest flyers. All local ecological projects are grist for the writer's mill. He should verify the facts and then prepare his manuscript, being sure to show how this plan or that demonstration is of local importance.

In choosing your first topic, try to find a project upon which action can be taken relatively quickly and which could conceivably produce immediate and visible results. If you can get results, you will have a supportive group ready and willing to back you on your next effort. A simple project worth writing about would be the removal of the junk from a stream near your town. A park might be created there and perhaps eventually swimming facilities. The potential for recreational activities is often overlooked—until a writer alerts the public to this potential. Even if so small a thing as a pond which is little more than a puddle is tidied up and used for skating, progress has been made and is demonstrable. The larger the flow of water in a stream, the greater its potential. The creation of parks along a river bank may lead to the installation of a swimming pool—and if the citizenry could be encouraged to *see* what has to be filtered out before the water is fit for swimming, it might start a crusade to clean out the sewage at the source rather than filtering it out downstream.

Any unique feature in the natural surroundings of your home should make a good topic. People are interested in nature. Perhaps there is something unusual in the geology, vegetation, or animal life in your vicinity which the right kind of ecological management could improve, preserve, and use as a means of attracting tourists. Tourism is an extremely important industry. Stone Harbor and Cape May Point, New Jersey, have added greatly to their tourist income by the creation of sanctuaries for birds. Cape May Point is now world famous as the focal point of the southbound Atlantic Flyway, and Stone Harbor has a nesting ground for herons set aside as a sanctuary. Pacific Grove, California, has become famous for its butterfly trees where the monarch butterfly rests in thousands (*National Geographic,* April, 1963), while

San Juan Capistrano and its returning swallows have become a worldwide symbol of dependability. Look hard. And look close to home.

Ecology Notebook

Every person interested in writing about ecology should keep some kind of ecology notebook. In it he should file away newspaper clippings, letters to magazine and newspaper editors, pictures, advertisements, announcements of meetings, and other ecological data. He can also jot down whatever ideas he has from time to time that concern the management of his particular environment or of the environment in general.

It is a good idea to include in such a notebook a "grievance list," that is, a list of examples of environmental abuse. From this list will come possible topics for papers. More important, once you begin to write notes about environmental problems that annoy you, you may well decide to do something about them. And perhaps you will start by writing—to newspapers, congressmen, state senators, town planning boards, corporations, clubs, and responsible citizens.

The ecology notebook can become a kind of personal resources book which will be useful when writing on ecological subjects. It is a good idea, for example, to jot down the addresses of societies and groups actively fighting environmental abuse. Often newspaper stories end with addresses to which you can write for relevant pamphlets and statistics. When the time comes to write, either on your own to a newspaper or a congressman, or as an assignment in a course, you will have a wealth of materials to call upon.

You may also wish to record in your notebook some dramatic statistics and facts of general ecological importance, such as those with which we opened Chapter One. The larger principles of environmental abuse can be used to support your discussion of specific local problems. You will—without conscious censorship—put in your notes the items which really interest you, and the sum of your notes will inevitably be sincere. Sincerity is a primary requisite of good ecological writing.

Consider the following as an example of how you might make entries in your ecology notebook:

> Came home on the subway today and realized how dirty my shirt was. Decided to discuss the dirt problem with some people in the next block; he's a biologist and she's a chemist. Amazed to hear their explanation: One day or less in most cities will soil a shirt,

blouse, socks, stockings, etc., often to the point that the clothing is unwearable. The clothing is washed, using water. Detergents are added, creating additional problems in the processing of the sewage, especially when phosphates are poured into the streams. Electricity is consumed, utilizing fuel and causing thermal pollution. And then the power plant belches out smoke and makes my shirts even dirtier. Clothing which has to be dry-cleaned causes a similar cycle. Dirt costs. I'll have to look into this problem further.

Read today that highways now cover with concrete an area the size of Massachusetts, Connecticut, Vermont, Rhode Island, and Delaware, and we are continuing to pave at the rate of more than an acre a minute. Of course, those are small states, and the papers are trying to be dramatic. Still, an acre a minute. . . .

Paul Ehrlich said somewhere that every American child is 50 times more of a burden on the environment than each East Indian child. Must check source.

Interesting television show last night. Apparently the Aswan Dam in Egypt is backing up and slowing the Nile. Sand bars have stopped the buildup of the Delta, and the Mediterranean is flooding it. A million acres of fertile land have disappeared under salt water, and the sardine industry is almost destroyed, with the result that the people are moving into the cities and extending the slums of Alexandria and Cairo. Wonder if there is anything parallel happening around here. What's making more people crowd into our neighborhood?

I can't get emotional about fish, but today I read that 150 million fish died of pollution this year! Swordfish are off the market now. I guess I should eat only little fish from now on: the bigger fish have eaten so many smaller fish that they are that much more contaminated. Mercury is the worst thing they have in them.

As I sit looking out the window, I'm thinking about the geranium on the windowsill. And on all the other windowsills. How sad it would be if we didn't have green growing things.

By keeping a journal in this way, jotting down ecological news items, personal thoughts, memorable statistics, or bits of philosophy, you will have a large collection of resources and hopefully some good original thoughts at hand to work into your paper.

Be Original, Be Relevant

In selecting your topic it is important to try to be original. To write about the same ecological problem that everyone else is writing about is a relatively inefficient use of time and energy. Try to write about what is important and what has not been previously discussed, or at

least not discussed at any great length, or from a particular point of view. As we have suggested, the grass roots level is the best place for the ecological writer to start. As his techniques improve, he may move on to the larger national problems, but there will be tremendous competition. At the local level the writer can be honestly emotionally involved, and sincere conviction tends to shine through even inept writing. Sincerity counts in moving human beings, and originality helps in getting their attention. An astonishing number of people whom one would expect to be well informed about what is going on in the world are not. Earth-shaking news never reaches them; some because of a deep involvement with personal problems, some because they have withdrawn from reality, and some for many other reasons. Yet these are the very people who need to be reached, and original thinking is most likely to bring them to an awareness of the ecological crisis.

In a curious way, it is sometimes easier to be original than to be relevant. What makes ecological writing relevant is its appeal to real citizens with real problems, such as the poisoning of a stream or the lining of gutters with broken glass. Sometimes the writer needs to work toward legislation at the same time that he is trying to influence someone or some group toward action. No polluter is going to clean up a situation for which he is fined one hundred dollars a day if it is going to cost him five hundred dollars a day to do so. That is why the writer must also work toward legislation that will close down polluting factories, storage areas, cleaning plants, car washes, and other businesses which continue to ignore fines and warnings.

Being relevant is putting at least some fear in the minds of your readers. The man who can imagine a distant benefit accruing to him as a result of a shutdown in a factory where he works is a rare man indeed. He will need to be shown by *facts and figures* that there will be some material advantage to him, at once or later. If later, he can be shown that eventually a particular mess or problem will have to be cleared up by the installation of expensive equipment and that the longer the action is delayed the greater the eventual cost to him (often through taxes) will be. That kind of information is relevant.

In summary, then, conscientiously keep an ecology notebook, making particular efforts to jot down all of your original ideas as well as summaries of pertinent newspaper and magazine articles. Select a topic close to home and present it with freshness and originality, while at the same time trying to appeal to your audience by emphasizing what is most immediate or relevant in your subject matter.

3 RESTRICTING AND DEVELOPING THE TOPIC

Restricting the Topic

When you have chosen your topic, you then need to restrict it, that is, limit its scope to a manageable size. Manageability generally is a function of the length of the paper you wish to write, and also of the research resources available to you in the library and other places. Don't try to include everything that can be written on the subject in your paper. Sharpen your focus. Make sure your goal is realizable, given your abilities, your resources, your interest, and your time.

Developing the central idea you have decided on is of crucial importance when writing on ecological topics. If you take your reader on too many detours (digressions), he may not derive the full importance of the one point you most want to make. At the outset, then, decide on your central idea, determine the time and the place of what you will describe, and determine as sharply as you can your own attitude toward your topic. When you are considering ways in which to restrict your topic, try to be specific.

Let's assume that you have been asked to write a 1,000-word essay on some kind of ecological topic. As we discussed in the last chapter, it is good to look for your subject close to home. If it is one of genuine interest to you, you probably have already jotted down some ideas about it in your personal resources ecology notebook. In any case, once you have the topic clearly in mind it is time for what most composition teachers now refer to as *brainstorming,* a frantic kind of energetic, prewriting activity.

Brainstorming

When you are brainstorming to get ready to write a paper, you are involved in free association, letting one thing in your mind lead to something else. You try to think of all of the ideas that seem relevant to your main topic in one way or another. You make a list of as many of them as you can, and then you try to organize them into a coherent outline.

The key to successful brainstorming is not simply to play the game of mental association with great concentration, but to keep asking yourself, while you are playing it, whether the particular idea you have at any moment can be logically included in your paper. In other words, if in the midst of your brainstorming you begin to realize that your ideas are degenerating, simply stop, wait awhile, and start again. Most people do tire if they force themselves to think too hard and too long.

Pierce Library
Eastern Oregon University
1410 L Avenue
La Grande, OR 97850

Let's assume, as an example, that you have decided to write a paper about the dead fish in the Great Lakes. First, you might say to yourself, "Actually, the only Great Lake I am familiar with is Lake Michigan, so I had better restrict my discussion to it." Your topic is now the dead fish in Lake Michigan. As you think about your topic further, you realize that the only dead fish you have ever seen in Lake Michigan are alewives. Thus you decide that you are going to write a paper about the dead alewives in Lake Michigan. So far your brainstorming has been directed toward restricting your topic, a manageable topic that you might be able to discuss adequately (or at least provocatively) in three or four pages.

Since you have selected the alewife, you decide to look it up in the dictionary. You discover (in *The Random House Dictionary,* College Edition) that the alewife is a "North American fish, *Pomolobus* (or *Alosa) pseudoharengus,* resembling a small shad." Just for the pleasure of it, and since your eyes are there anyway, you also read the second definition listed for alewife, "a woman who owns or operates an alehouse." First you laugh, and then you start wondering what the fish and the woman have in common. You ask your roommate, but he doesn't know what you are talking about. You call him a *pseudoharengus.* He says to leave him alone because he is trying to write an anthropology paper and doesn't want to be bothered. The next day you ask your English teacher, but he doesn't know. He feels guilty. He's beginning to wonder whether English teachers should be organizing their writing courses around ecology. You return to the dictionary and look up shad. You discover that it is a kind of herring that migrates upstream to spawn. Could that have something to do with it? All day long you walk around thinking about the word "alewife." Finally you look it up in an encyclopedia in the library. You discover that the fish alewife possibly got its name from the Indian name, *aloofe.*

One thing is now certain: You have been wasting too much time thinking about the name "alewife" and you have not been thinking enough about the paper you have to write. Serious brainstorming must now begin. In condensed form, your thinking might go as follows:

> Alewife: Camping last summer at Muskegon on shore of Lake Michigan, swimming, bumping into dead fish. Yuk! Pollution. Filthy water. Why bother? Because it's very hot, that's why. A hot day and stinking fish all around. Alewives, swimming their way into the big lakes and getting locked in where they can't live very long. I remember someone told me that the Coho salmon had been introduced into Michigan waters to eat the alewives and prevent them from getting as far as Lake Michigan. But I remember, too, a picture in a Detroit newspaper—I think the Detroit *Free Press*; I'll have to check that—of fishermen, real sport fishermen, clubbing Coho

salmon with baseball bats as the salmon traveled upstream . . .

And so it goes. Your mind simply keeps turning over related ideas, recalling, inventing, puzzling, and all the time trying to come back to the alewives. Slowly you realize that it shouldn't be too difficult to write a thousand words on your topic. Now you are ready to make an outline.

The Outline

When you have finished brainstorming, that is, when you feel ready to plan your paper, not just think about it, you should make what is called a "scratch outline." This is simply a list of the ideas you have generated through brainstorming. Then you can proceed to either a topic outline or a sentence outline. A topic outline uses single words, phrases, and clauses to indicate ideas. Only the thesis is actually written as a complete sentence. In a sentence outline, however, all parts of the outline are expressed as complete sentences. Usually the sentence outline is more productive because it forces you to think through your ideas and thus get a fuller sense of how you might write them in your final paper. By definition, a sentence outline, as opposed to a topic outline, has to be more detailed; it requires you to think ahead about particular points or examples you may include. By writing a sentence outline, you save yourself time and work later on. Obviously there is nothing wrong with making first a scratch outline, then a topic outline, and finally a sentence outline. But in our view it makes more sense to proceed directly to the sentence outline.

In organizing your outline, we suggest the following procedure:
1. Precede your outline with a clear statement of your central idea.
2. Divide your notes and ideas into related groups and arrange the groups in a meaningful order, always keeping your central idea in focus. Ask yourself whether every idea is relevant and necessary to your purpose.
3. Decide on the main divisions and subdivisions of your topic, writing a descriptive statement for each. Try to provide at least two subdivisions for each divided entry. If you can think of only one subdivision, it is usually better not to have any subdivisions at all for that particular entry.
4. It is a convention of outlining to number the different divisions using the following symbols: Roman numerals for the main parts, capital letters for supporting ideas, Arabic numerals for subdivisions of the supporting ideas, and lower case letters for further subdivisions. Usually, but not necessarily, the main, Roman numeral divisions will correspond to paragraphs.

24 Ecology: A Writer's Handbook

The "skeleton" of such an outline is shown below. Both topic outlines and sentence outlines follow this format.

I. First main point.
   A. Supporting idea for I.
      1. Supporting idea for A.
      2. Supporting idea for A.
   B. Supporting idea for I.
      1. Supporting idea for B.
      2. Supporting idea for B.
         a. Supporting material for 2.
         b. Supporting material for 2.
II. Second main point (division).
   A. Supporting idea for II.
   B. Supporting idea for II.
      1. Supporting idea for B.
      2. Supporting idea for B.
   C. Supporting idea for II.
      1. Supporting idea for C.
      2. Supporting idea for C.
III. Third main point (division).

. . .

Now let's make a sentence outline for a paper on the dead alewives in Lake Michigan.

Summary: There are many dead alewives in Lake Michigan, and although some measures are being taken to prevent them from being there, further efforts are needed.

Title: Dead Alewives in Lake Michigan.

I. There are thousands of dead alewives in Lake Michigan.
   A. I know this first-hand because I encountered them while swimming in Lake Michigan last year.
      1. Many bathers are protesting loudly.
      2. Lake Erie is already unswimmable; must Lake Michigan die too?
   B. The alewife problem is typical of environmental abuse.
      1. The alewives swim into Lake Michigan and are locked in.
      2. They die in Lake Michigan because the water is so badly polluted.
         a. One can see signs of raw sewage in Lake Michigan.
         b. There are no signs of life in Lake Erie.
II. Conservationists in Michigan are having some success in dealing with this problem.
   A. Numerous government-sponsored studies have been made of the dead alewives in Lake Michigan.
   B. The state government is actively trying to reduce the number of dead alewives.
      1. Some industrial concerns responsible for polluting the lake have been penalized.

2. The largest part of the government's attack on the problem has been introducing the Coho salmon into Michigan waters.
   a. The Coho salmon is a large, beautiful fish.
   b. It eats many of the alewives before they have had a chance to swim into Lake Michigan.
III. The Coho salmon is an ideal choice for dealing with the alewife problem, but it is not doing the full job that is necessary.
  A. There are numerous kinds of fish migrations, and we are still in the process of learning about them.
    1. The Michigan government knows that the Coho salmon will move in such a pattern as to eat a large number of the alewives before they reach the lake.
    2. Studies being made of the herring industry in Canada may help.
  B. Controlling fish migrations and populations is becoming more scientific, particularly with regard to controlling thermal and salinity changes in different waters.
  C. Unfortunately, the Coho salmon is attractive to sport fishermen, and many of the salmon are caught before they have had a chance to eat the alewives.
    1. Couldn't the government tighten the laws regarding the fishing limit (number of fish per fisherman) on Coho salmon? Could certain waters be restricted from fishing for salmon?
    2. In the Detroit *Free Press* sport fishermen were pictured clubbing Coho salmon with baseball bats as the fish swam upstream. (Put photo in an appendix.)
IV. It is time to increase the fight against the dead alewife problem in Michigan.
  A. Though changing the game laws on catching Coho salmon will help to some extent, it will not be enough.
  B. Strong practical measures, involving penalties and taxes, must be taken against the industrial concerns most responsible for polluting the lake.
    1. Statistics (cited with full references) show the amount of pollution in Lake Michigan.
    2. Comparative statistics on pollution in the other Great Lakes, particularly Lake Erie, indicate the seriousness of the Lake Michigan pollution.
  C. The companies responsible for the pollution of Lake Michigan are: (companies identified by name and kind of pollution).
  D. By simultaneously limiting the Coho salmon season and increasing the penalties on polluters of the lake, Michigan ought to have a good chance to solve the problem of dead alewives in Lake Michigan.

It is relatively clear throughout the outline that the writer has his main concern in steady focus. There are no unnecessary digressions. The outline ought to serve as a good foundation on which to build a persuasive essay.

Supporting Evidence

Gathering supporting evidence is important in all kinds of writing, but particularly in ecological writing. When you are writing about an ecological problem, it is very important to have your figures and ideas correct. Otherwise, in some readers' minds, you will only be attacking a paper tiger. If you are going to write about pollution, you must be able to cite evidence, to *prove* that there is pollution where you say there is. The testimony of witnesses, records, documents, science writers for the local press, scholars in relevant fields, officers of garden clubs and sports clubs, publications of the local conservation society, plus of course all the wealth of information in government and public libraries—all these constitute authoritative sources for obtaining your supporting evidence.

Furnishing evidence, providing proof of various aspects of your topic, is the all-important final step to take between writing your sentence outline and writing your first draft. Look over your sentence outline. Where would supporting evidence be most helpful? What points might not convince a reader without supporting evidence? Looking over the sample sentence outline of a paper to be written on the dead alewives in Lake Michigan, you might decide to gather supporting evidence in the following places:

I.A. Testimony of other qualified people as to the estimated number of dead alewives in Like Michigan. The state department of fisheries ought to have precise figures.

I.B.2. a and b. Supporting evidence (e.g., comparative figures for other lakes) for these assertions would strengthen your argument and perhaps dramatize the problem effectively.

II.A. Cite some of the government studies.

II.B.1. Cite some typical penalties.

III.A.2. Cite some of the studies.

III.C.1. Summarize the present laws; perhaps explain the fishing season.

IV.B.1. and 2. Use statistics.

IV.C. Evidence here would list the companies and their businesses, kinds of pollution, waste, etc., located on the shores of Lake Michigan, particularly the worst polluters near Chicago or Benton Harbor.

By taking a red pencil and reviewing your sentence outline with a particular concern for where in the paper supporting evidence is needed, you will be saving yourself time later on when writing the paper. Your evidence should be so "solid" that it provides obvious proof of your statements and shows your argument as a whole to be entirely convincing. It is easy to counter opinion with another opinion, but when your opinion is supported by a great deal of concrete evidence, there is not much room for anyone to attack it or to doubt the authenticity of your argument. Also, in terms of good ecotactics, if you are going to attack someone or some group for bad ecological management, it is always wise to consider ahead of time all the ramifications of your argument. You will probably be stepping on a good many toes: if you can create a climate of consultation, of fact-digging, you will leave less room for a counterattack from those you are attacking. Therefore, consider all possible sources of further evidence. Ferret out facts. Look through town records. Inspect local facilities yourself if they are being questioned. Interview people who know something about your topic. Extreme tact can produce excellent results and a great deal of information, since most people are willing to talk when someone indicates a genuine interest. Consult journals and articles in the press as well as the better books for the lay reader and texts on various aspects of ecology (see the Bibliography of this book). Words in print convey considerable authority. Also, there is nothing wrong with citing legislation of any kind—if in some way it constitutes a kind of supporting evidence. For example, if you are arguing that not enough attention has been given to a particular problem, you may well cite the voting record of certain candidates or elected officials on environmental problems. Such facts can be very helpful, especially if you are writing near election time.

In marshaling evidence, it is also a good idea to take photographs and assemble them as an appendix to your paper. Pictures of dumps, sewer outfalls, junk-filled streams, cluttered ponds, and belching smokestacks are of material value in getting the message across. (Incidentally, a valuable technique is to place the masthead of the local paper somewhere in the picture where it can be read, even if only with a magnifying glass; this establishes the date in case one of your opponents claims that the mess was cleared up long ago.)

The ecological writer must be prepared to take a great deal of time in the pursuit of such illustrative materials. (Eastman Kodak has an excellent book on this. See Eastman in the Bibliography.) It is extremely important that he compile extensive evidence; even if he is guilty of the charge of using a baseball bat to kill a flea, at least the flea will die, and we may all live longer.

However, remember not to try to sound like more of an expert than you are. Let your experts present their own testimony (through quotation). And don't talk in overly technical terms that a layman will have difficulty understanding; parts per million and milligrams per square meter are not very meaningful to the average reader.

Finally, when you present your supporting evidence, let the figures like the experts speak for themselves. Facts and figures are dramatic in themselves and don't need the cosmetic embellishment of rhetoric. In gathering data, remember Rachel Carson's *Silent Spring* and the efforts made to discredit it; the public is so conditioned to shrill voices that it often no longer hears them—sometimes it helps to whisper! This does not mean that a documented essay must be cold, formal, or entirely impersonal, but rather that in a documented essay the facts and figures themselves will be dramatic, and the writer's own tone should thus be even and controlled. The whole point in gathering evidence is precisely to free the writer from the temptation of falsely dramatic performances.

4
TECHNIQUE, STYLE AND PERSUASION

When you have selected, restricted, and outlined your topic, gathered all your evidence, and feel ready to write, you will be involved in a number of *selecting* activities: What method of discourse shall I use? What audience shall I address my paper to? What kind of words would be most appropriate in addressing that audience? What combination of ingredients will most successfully persuade the audience to accept my point of view? The purpose of this chapter is to introduce information about writing that will enable you to answer these questions.

Main Techniques of Composition

There are four traditional methods of discourse: description, narration, argumentation, and exposition. You usually select one or more of these modes, according to the purpose you have for writing; so let's consider the differences among them.

DESCRIPTION

The purpose of description is the presentation of a picture in words. To describe something is to *depict* (think "picture") it. Your picture may be made with information and facts, resulting in *objective description*:

> The lake in question is twelve miles long and two miles wide. Its average depth is forty feet, though at the center it is eighty-five feet. The shore of the lake is lined with timber, mostly oak and pine. Recent water samples indicate that the lake is becoming increasingly polluted.

Or your picture may be mostly imaginative, which we refer to as *subjective description*:

> The lake, which is long and narrow, looks like a long band of ribbon in the moonlight. What a tragedy that the package beneath the ribbon is foul and polluted.

In general, and particularly when writing an essay on an ecological problem, you would not limit yourself to description but would combine description with one or more of the other modes of writing.

NARRATION

The telling of actual events is called narration. It is a record of occurrences, generally presented in chronological order. When you narrate,

34   Ecology: A Writer's Handbook

you tell A, then B, then C, and so on, usually to recount and explain. Sometimes narration is imaginative (as in narrative fiction), but in ecological writing you almost always arrange true events in their chronological order so that your audience will be able to understand the resultant situation. The following passage suggests how narration might be used:

> A few years ago people started throwing their soda bottles, beer cans, and cigarettes off the bridge by the high school into Tony's Brook. After a while you could see the refuse starting to build up, and by the end of the year you could find refuse about a mile downstream. Last month there was refuse all the way down by the new recreation park, not just a few things, but large amounts. The ducks no longer land on the water because the water is foul. Children don't enjoy playing by the brook in the park because the water smells. What a sad story, that the initial negligence of a few dozen people soon can deprive hundreds of children of the pleasures of feeding the ducks and playing by a clear brook.

By narrating this "sad story," the writer is making it clear to his audience that the events occurred in a progression. There is clear evidence of cause and effect here. The orderliness is useful in showing that certain acts have eventual consequences, and that something done here affects something over there. Ecology is, after all, a story about life, and about how various aspects of nature are interrelated. Narration can be very effective in demonstrating ecological mismanagement.

ARGUMENTATION

The goal of argumentation is to convince the audience of a point or a position. When you argue, you are trying to persuade (and often to justify). A "good" argument, technically at least, is a valid one, which is to say a logical one. When you argue you do not need to have the one and only "truth" on your side, but you do need to be tightly logical, rigorously fair-minded, and at the same time persuasive.

An example of argumentative writing might look like this:

> It has been known for some time that many of the industrial factories on the shores of Lake Michigan have been dumping their waste chemicals into the lake. This dumping has been going on for many years. You can't expect the results not to show up eventually.
> Last year there were large numbers of dead alewives in Lake Michigan. Swimmers throughout the Traverse City and Muskegon sections of the lake complained steadily about bumping into dead fish every few minutes.
> All the chemical concerns are still dumping their waste products

into the lake. These chemical wastes are killing the alewives, and the dead alewives are bothering the swimmers. Thus it follows that the industrial concerns are the ones responsible for ruining the swimming in Lake Michigan.

This argument is making use of the most useful (and most common) construction of logic, known as the *syllogism*. A syllogism, as most students who have studied Venn diagrams and set theory know, can be represented in the following way:

                   All A is B.
                   All B is C.

Therefore:

                   All A is C.

In the prose example of argumentation presented above, the same syllogistic pattern is used to show that the industrial polluters are responsible for ruining the swimming in Lake Michigan. In any case, in argumentative writing, your main goal is to persuade, to win over your audience, and we'll have more to say about that in this chapter. But first, let's consider the fourth mode of discourse.

EXPOSITION

Exposition is *explanation*; in expository writing you are presenting sufficient information to explain something. Directions in books, definitions in dictionaries, and most textbooks are written in expository prose. Since most of the writing assignments in college are analytical, requiring explanation of various ideas, experiments, or events, most of them are also expository. And most expository writing relies on at least some factual materials:

> I'm not sure why I first began to loathe cigarettes, but I think it surely had to do with growing up in a smoke-filled house. Not only my parents smoked, but all their friends did too. When they came in to kiss me good night, their clothing and their hair reeked of stale tobacco. When I first tried smoking I coughed, as I guess most people do the first time they try it, and was hard-pressed to understand why people smoke. Now it has become clear that there is a relation between smoking and lung cancer, heart disease, and various other illnesses. Medical research shows that a heavy smoker shortens his life by 6 minutes for every cigarette he smokes. If you are a heavy smoker, your chances of dying between 25 and 65 years of age are twice as great as those of the nonsmoker. No, I'm not sure why I never began smoking, but I'm certainly glad I didn't.

In this exposition, the writer is combining autobiographical reportage with explanation; he explains not only his own nonsmoking but also the risks in smoking. In expository writing you are giving an account of something, making it clear to an audience, usually for a reason—in this case, to suggest to the audience that it is unwise to smoke.

While each of the four modes of discourse has its particular features and intention, more often than not you will be combining several modes. In ecological writing, usually exposition will be the dominant mode in the combination, and you will be writing "expository narration," "expository argumentation," or similar combinations. That is, most of the time you will be writing expository prose; but within this main mode, you will be varying your method of presentation according to your audience and your intentions, and sometimes shifting to description, narration, and argumentation.

Persuasion and the Audience

If you are going to be a forceful, persuasive, and convincing writer, you will need a sense of the audience for which you are writing. Remember that even if you are writing a paper for a course, you are not writing simply for your teacher. Is your assumed audience politically affiliated? highly educated? upper-middle-class? a group of hippies? Since every writer wants to have an impact on his audience, he must have its general nature in mind as he *starts* to write.

The all-important relation between style and audience is based on *appropriateness.* You cannot rely upon colloquial words and expressions when writing for an audience which *expects* a formal analysis. The language of formal analysis would, in contrast, probably be inappropriate for "reaching" an audience consisting entirely of teen-agers. A writer wants to be effective for (have the desired impact on) a *particular* audience. This implies that a writer must have some understanding of his audience's experiences and attitudes.

Therefore, think of your intentions: What are you trying to accomplish by writing this essay? What kind of response do you hope to evoke? Do you wish to shock your reader? flatter him? enlist his support? objectively increase his understanding of some particular subject?

Consider the opening of a speech by Bobby Seale about Huey P. Newton, a speech intended not only to defend Newton but to explain to black people the perspective on their oppressed conditions which Newton provided:

> Brothers and sisters, tonight I want to have the chance to tell you in large mass something about Brother Huey P. Newton; a black man that I've been knowing for about eight years. . . . To explain to you who Brother Huey P. Newton is in his soul. I've got to explain

to you about *your* soul, *your* needs, your political desires and needs, because that is *Huey's* soul.*

Bobby Seale obviously knows and relates quickly to his all-black audience.

Sometimes a writer not only identifies and utilizes his audience, but also uses the narrative technique of placing his reader in a certain situation in order to facilitate the description of it. Generally this calls for the direct use of "you," though it does not need to. The twentieth-century essayist Gilbert Highet begins a discussion of "kitsch" (a Russian-derived term, meaning, as Highet explains, "anything that took a lot of trouble to make and is quite hideous") in the world of books by asking the reader to recall or imagine his wanderings through antique shops:

> If you have ever passed an hour wandering through an antique shop (not looking for anything, exactly, but simply looking), you must have noticed how your taste gradually grows numb, and then —if you stay—becomes perverted. You begin to see unsuspected charm in those hideous pictures of plump girls fondling pigeons, you develop a psychopathic desire for spinning wheels and cobblers' benches, you are apt to pay good money for a bronze statuette of Otto von Bismarck, with a metal hand inside a metal frock coat and metal pouches under his metallic eyes. As soon as you take the things home, you realize that they are revolting. And yet they have a sort of horrible authority; you don't like them; you know how awful they are; but it is a tremendous effort to drop them in the garbage where they belong.†

It is interesting to observe the way in which Highet's writing style— informal, chatty, wry, humorous in tone—relates closely to the situation in which he is picturing his audience. When he writes the phrase, "to drop them in the garbage," he knows that his reader is smiling, enjoying not only the account but the phrase. And this is what the student writer should always be trying to accomplish—anticipation of the audience's response both to the subject matter and to the *stylistic presentation* of the subject matter. If you want to convince someone that he should do something about a particular ecological problem, you start by interesting him in the way you are writing about that problem.

LANGUAGE AS A TOOL OF PERSUASION

When thinking about the kind of language you wish to use, you must make choices between significantly different alternatives. Your diction

---

* Bobby Seale, speech printed in *The Black Panther* (March 16, 1968).

† Gilbert Highet, "Kitsch," *A Clerk of Oxenford: Essays on Literature and Life* (New York: Oxford University Press, 1954), p. 210.

(choice of words) will need to be either mostly formal or mostly informal, depending of course on your audience, subject matter, and intention. The words you select should be of a like nature, that is, should seem to be going well together. If you mix the levels of formality in a radical manner you will usually jar your reader's nerves. You want your reader to feel that you are comfortable about your expressions, that you feel satisfied with them, and this in turn requires that you make the reader feel comfortable. You do that, in part, by satisfying his expectations. If he has read the first page of your paper (say, a relatively scholarly analysis of an urban ecological problem), he expects the next page, and the next, to be written in the same kind of language.

A writer has to determine at the outset what kind of language is *most appropriate* for discussing a particular subject. However, language is not rigid, and a certain amount of creative flexibility is desirable. In the end, the amount of departure from the general level of formality in an essay depends on one's own taste and on how well one knows (or thinks he knows) his audience.

STYLE

Style is the selection and arrangement of words; it includes diction, consideration of audience, tone, intention, and personality. Be careful to think about every word you use in the contexts of its suitability to the subject, its suitability to the rest of the paragraph and sometimes even to the rest of the paper, its tone, and its precision or superiority over other possible words and expressions with roughly the same meaning.

One of the most important elements of style is tone, or the writer's attitude toward his material. Your writing should make clear to your reader *from the very beginning* just what your tone or attitude *is*. Your tone will indicate *degree of subjectivity*, which should be determined naturally by your topic and the purpose of your essay.

In summary, all general considerations of your audience, tone, language, and degree of subjectivity are, or should be, interlocking. You should realize that everything in writing has to do with everything else. This is why it is useful to think of writing as a decision-making process. Do you write A or B? If you write A, should you let it stand by itself, or would it be stronger if followed by C or D? or perhaps by both C and D? When you use one verb and not another, will you be altering your tone? The questions keep coming. Different kinds of writing inevitably bring out different kinds of questions, but you will always find it necessary to be asking and answering them.

WRITING FOR THE GENERAL PUBLIC

If you are writing an ecological essay for the general public, there are certain factors to bear in mind. Your tone should not be unduly authoritarian. Most of the twenty-eight essays in the widely read book *Ecotactics* (see Mitchell in the Bibliography) are conversational and informal—even though some very startling facts are being presented. You should not use so many figures and facts that you confuse the average reader. If you use a quotation, try to find one that is not frayed with overuse. However, if a popular quotation precisely fits your train of thought and is an excellent way of introducing your theme, do use it. People often prefer to move into unfamiliar territory on an at least semifamiliar road.

An ecological essay written for the general public should have a core of hard facts. Facts are necessary to inform the audience, arouse their sentiment, and stimulate action. For the general audience, you should not assume too much advance knowledge of the subject. Thus, clarity of presentation and a logical progression of ideas from start to finish are very important.

When writing for the general public it is good not only to make use of narration, but to dramatize the topic by casting it into story form. If you have ever attended an old-time movie theater, or a simulated one, you will remember how the organist or pianist played throughout the production. There were certain set compositions which he used to underline the plot. The children's operetta *Peter and the Wolf* is a good example of the same technique, and there would be considerable merit in listening to it before writing an ecological essay for the general public. Mentally select the villain and the hero and use varied styles as you write about them. The villains are the perpetrators of pollution. Facts about the problems should be presented in short, pointed statements, backed up by authorities. Your hero is your audience. He can kill the dragon; he can shoot the villain; he can chop the giant's head off. In addressing him your theme music should be hopeful, friendly, understanding, inspirational.

Now devise your "plot." What can your hero do about the villain? How can he do it? Be specific. There is little point in exhorting the reader to insist upon a clean and decent environment unless you can tell him the means by which he can make his insistence effective. First you must discover who is responsible for the villainy, and this is not always easy. Who is doing the polluting? Where is it being done? When is it being done? Why is it being done?

After you have pointed out the enemy and shown why he is to be judged the enemy, provide your audience with weapons. There are various ways this can be accomplished. A thorough perusal of town ordinances and state laws will often turn up a forgotten legal weapon. An important case in point is the series of Rivers and Harbors acts culminating in the act of 1899 which has enabled conservationists to fight polluters legally. Such research takes time, but it can be very rewarding. Also, there are legal conservation consultants connected with large organizations; direct your readers to them. They have less familiar resources and techniques at their finger tips.

It is extremely difficult to write about matters that are of intense personal concern without slipping into the first person ("I"). All of us do it. Nevertheless, avoid it when possible. Unless you have considerable stature, the public remains unimpressed by what look like ego-parades in print.

A few words of caution about writing for any audience, but most particularly for the general reading public: Be careful not to make a damaging statement about an individual or organization unless you are absolutely certain the statement is true and that you can prove it. You will be sued for libel if you are careless in this regard.

ADDRESSING ECOLOGICALLY KNOWLEDGEABLE AUDIENCES

Let's assume now that you are addressing an audience which is relatively well-informed in matters of ecology, for example, in a speech to a conservation group or a letter to a board of governors of a nature club. You will want this group to respect you for what you say. They will not be as easily impressed by the general statements you might include in an essay written for the general public. The well-informed audience will want more facts and figures, and more precise and technical ("insider") language than the general public will want. It is important when writing for a well-informed audience to be unusually careful about your sources of information. Certain groups (especially certain civic-minded groups) have their private "hate" lists of several publications. Try to know what they are, and then avoid them. Obviously, the reverse can be true. If one of the people in the group to which you are writing himself writes for a certain publication in the field, it might well be worth going out of your way to quote something from that publication. It should be clear to you by now that consideration of audience is itself a central aspect of ecotactics.

THE POTENTIALLY HOSTILE AUDIENCE

In addition to thinking of the differences between the general audience and the ecologically well-informed audience, you might do well to think of the potentially hostile audience. For example, when presenting radical measures to a conservative audience, be particularly muted in your writing. Don't scream at your readers. They won't listen. Be cool and reasonable, for that is how they think of themselves. And don't urge any hasty action. Give your readers time to consider the problem. Conservatives are suspicious of rapid change. It is better not to *tell* them too much, but rather to use a calm attitude that seems to ask, "Do you think there is an idea here? Doesn't this make sense to you?" Similarly, you would alter your approach if writing to a liberal audience, filled with action-oriented overnight-change types. Rally them to arms. Paint the villain in broad strokes. Give them lots of ammunition—from the stark facts to ways in which they can immediately be helpful.

WRITING TO THE GOVERNMENT

If you are writing to a governing body, say a town board of governors, it is valid and wise to state your qualifications or at least explain who you are, since many such letters are by law read aloud at town meetings, public hearings, and similar functions. Usually, few members of this audience will know who you are. If you are writing in behalf of a group of people, that fact should be mentioned in your first remarks. Again, supportive opinion from recognized specialists or those whose titles would make the hearer reasonably sure that they are qualified will add considerable weight to your argument.

Unhappily, perhaps, many of the people you will write ecological letters to are in positions of power. If some of them seem fools to you, there is no need to let them discover that you think so. There is no point in making your readers hostile and defensive. Find *something* good that you can say about them and say it. They may do something constructive, even if only by accident. Make your suggestions to people in powerful positions with the implication that they are more knowledgeable than you are and probably themselves had thought of all of the things you are telling them. This may be somewhat hypocritical, but it masquerades under the guise of tact.

Two more notes of advice may help you persuade your audience. Never make sweeping emotional indictments, for there is almost inevitably an exception to be cited by your adversary. He will use that exception to destroy the validity of all the rest of your thesis, no matter how proper and justified it may be. Finally, sarcasm is a two-edged sword; eschew it as you would the devil. There is much merit in the old statement that nothing is more fatal than a difference of opinion as to what is funny. Senses of humor are variable and unpredictable. Be extremely careful, and think long and hard about the nature of your audience, before attempting to use humor or irony; it may very well backfire.

Persuasion in Different Literary Forms

Most of the ecological writing you do will take the form of either the brief 1,000-word essay or some kind of letter. We have thus far said more about essay writing than letter writing; it now seems time to redress the balance.

THE LETTER

Before you undertake to write a letter, you should of course do the brainstorming, scratch outlining, and sentence outlining that you would do before writing an essay. Sticking to the main point is as important in a letter as it is in an essay. In fact, it is probably safe to say that whether or not your letter will be influential depends on how effectively it makes its main point. The following advice about letter writing is included here not simply to save what some consider a dying art, but rather to provide you with the "correct" (and therefore attention-getting) format for letters.

When writing a letter there are six parts you must include:

1. The heading (your address, the date).
2. The inside address (the name and address of the person to whom the letter is addressed).
3. The salutation (opening greeting).
4. The main text of your letter.
5. The closing.
6. The signature.

Though of course you are most interested in point 4, if you want your audience to take your letter seriously, it must be typed in a standard format.

The heading is best presented in the upper right-hand corner:

145678 Riverside Drive
New York, New York 10025
July 8, 1971

Some writers prefer to indent each consecutive line, but in general the blocked (straight, even left margin) presentation is preferred.

Type the *inside address* in the same blocked manner as the heading:

Mr. John Garble, Public Relations Director
Allegheny Suds Co., Inc.
426 Mercer Drive
Detroit, Michigan 48100

The *salutation* (greeting) should be appropriate to the tone of your letter. Probably it is best to be as formal as you are irate. All of the following salutations are considered formal:

Dear Sir:
Dear Madam:
Gentlemen:

Sir: (most formal)
Dear Mr. President:
My dear Mr. President: (very formal)

The salutation should be even with the left-hand margin of the inside address. The inside address should be set four lines below the bottom of the heading, and the salutation should be set two lines below the bottom of the inside address.

The important point to remember about the *main text* of your letter is that it should all relate to the central point. Usually several paragraphs of medium length will do the job well, although sometimes you may want to write very short opening and closing paragraphs with one long middle paragraph. If there is little time to digress in a three-page essay, there is certainly little time to digress in a one-page letter. Don't wander. Don't be melodramatic. Don't be "cute." State your main concern clearly and forcefully. Each paragraph should be typed single-spaced (same as the heading and the inside address). There should be two lines between the salutation and your first paragraph. There should be two lines between each paragraph, and two lines between your final paragraph and your closing phrase.

The *closing* should also be consistent with the tone of the letter. Some relatively formal closings are:

> Yours respectfully,
> Respectfully yours,
> Very respectfully yours,

A little less formal are:

> Sincerely yours,
> Yours sincerely,

Friendlier are:

> Warmest personal regards,
> Cordially,
> Faithfully yours,

The *signature* should be signed by hand in ink. Underneath the signature it is polite to type your name, and, if you have one, your title.

Keeping these various points in mind, let's examine the following Sample Letter to a Corporation.

Sample Letter to a Corporation

145678 Riverside Drive
New York, New York 10025
July 8, 1971

Mr. John Garble, Public Relations Director
Allegheny Suds Co., Inc.
426 Mercer Drive
Detroit, Michigan 48100

Dear Sir:

    I am writing with regard to your company's recent television commercial shown at 9 P.M. Wednesday, July 7, 1971. It was a dreadful commercial and, in my opinion, did more to harm your company's image than it did to help it.

    In the first place, most ecology-minded citizens are already very annoyed with companies that persist in using phosphates in their detergents. Your commercial made it shockingly clear that you persist in thinking phosphates are necessary. It seems insane to advertise the fact that you are not a progressive company. If the only way you can get clothes clean is to use phosphates, you had better not advertise at all.

    In the second place, your attack on other detergent companies seemed in very bad taste. You ought to realize that most consumers have an affection for underdogs, and thus for you to launch a derogatory attack on your competition is, really, to help your enemy.

    In short, I urge you to get in touch immediately with your advertising agency (it may be better for you to switch than fight!) and request them to cancel further airing of that commercial. The combination of discussing phosphates and of viciously attacking your competition will result in a great deterioration in your sales.

Respectfully yours,

John L. Chlorine

John L. Chlorine, Chairman
Riverside Ecology Center

Notice that the tone of the Sample Letter to a Corporation is consistently belligerent. It is clear throughout that the writer was very annoyed with the company's television commercial, and that he was determined to let the company know just how annoyed he was and why. Notice that the writer does not wander from his main concern. Notice, too, that there are no veiled threats—just the simple, direct, strategic suggestion that people might feel sorry for the underdog competition and thus hurt Allegheny Suds' sales. It is always safe to assume, when writing to any company, that they are most sensitive about sales. Aim at the company's profit picture if you want to get good results.

Through a series of short paragraphs the author managed to cover a great deal. The letter is forceful, direct, clear, and ought to prove effective.

THE LONGER DOCUMENTED LETTER

As opposed to a brief letter to a corporation or to an editor, you may want to write a longer letter to a company, a legislative group, a senator, and so forth. The longer letter will contain more arguments and thus generally more data and documented material. Our Sample Longer Documented Letter, written by one of the authors to the Planning Board of Shelter Island, Long Island, may serve as an example of this kind of letter. Notice that the tone is argumentative throughout, but the writer always attempts to be polite. (See p. 47.)

LETTER TO THE EDITOR

When writing a letter to a newspaper or magazine you are writing for publication. You hope that your letter will appear in print and be read not simply by one public relations executive or one town commission, but by thousands of other citizens. They are citizens that you would like to win over to your side, to your point of view. Thus, it is extremely important to be as logical and as persuasive as you can. Every town newspaper has certain citizens who write so-called crank letters every time a bird looks at them cross-eyed. Nobody reads their letters in the newspapers very often. So, if you do start writing letters to the editor, don't overdo it. The burden is upon you to be forceful, not copious. Consider the writer's approach in our Sample Letter to the Editor (see p. 49).

The sample letter is polite and persuasive. It is written in a reasonable tone, though it has a definite point of view to deliver. The letter takes the pattern of contrast as the basis of its organization. The writer goes so far as to admit that it would be nice to have both A and B, but, being realistic and a taxpayer, he realizes there is probably only enough money for A or B, and, as he explains, B would be better.

Sample Longer Documented Letter

                                State of New Jersey
                                Montclair State College
                                Upper Montclair, New Jersey 07043
                                September 17, 1970

The Planning Board
Shelter Island
New York

Gentlemen:

    At the request of many deeply concerned citizens, I have made an investigation which may be of interest to you. I am Professor of Biology with a speciality in Environmental Science at Montclair State College. Mr. Kelland and Mr. Ramsdell are Professors of Geology here, and respected colleagues.

    Attached you will find a Xerox copy of a cross-section Schematic of Long Island. Shelter Island is in this geological formation. You will also find an editorial from yesterday's *New York Times.*

    You will note that the fresh ground water is floating on salt water. Obviously as it is pumped out, a salt water intrusion takes place. *The New York Times* of March 7, 1965, is the source of the map. Even there the intrusion had reached well past Bay Park and was progressing at a rate of one hundred feet a year. You will also note that the greatest extension of the intrusion is taking place on off-shore islands, similar to our Island, but more populated.

    At the postulated one hundred feet a year nibbling from all of our shore line, such an intrusion would take at the most five years to destroy all of our shore line waterfront property so far as potable water is concerned. Then we would be forced to put in a water system. Since I am not an engineer, I cannot estimate the cost of such a system to the taxpayers, but I am very much interested in that cost since I am a taxpayer. I should judge that it would run into the millions.

    Shelter Island has at no point an adequate sewage treatment plant. The HEIGHTS simply holds the sewage and treats its chemically, a primitive method and an inadequate one.

    The accepted distance in most places between a well and a cesspool is about one hundred and fifty feet. I would like to point out that Shelter Island is composed of very sandy soil; hence bacterial penetration is more rapid and travels further. What is more, our water is held in place over the salt water by the pressure created by the rainfall. The rainfall is inadequate to cope with very many additional wells.

    The bacteria from the cesspools, once they reach the ground water supply, will thus travel with alarming rapidity, not only through spotty selected areas, but through the whole water supply. If one defecates on the grass-covered heavy loam, there is less penetration of pollutants than if one defecates on a sponge, which is in effect what a cesspool on Shelter Island may be compared to.

Planning Board, Shelter Island   Page 2

The addition of many closely distributed cesspools particularly in low areas, such as Hay Beach and Sunny Slope, will foul the water supply for the entire Coecles Harbor area for a certainty, and will probably foul it for the whole island.

As a biologist, I am also interested in the hunting and fishing. I do not weep salt tears over the taking of a brace of duck for dinner, or the taking of fish for sport, and I am also very much interested in the scallop harvest. I am told that the harbor harvest in marine organisms is worth over two hundred thousand dollars per annum. This to the Hare Leggers, and I consider myself to be one, is not an inconsiderable sum. Cesspools on Sunny Slope will ruin the breeding grounds for duck and other game, and at Hay Beach and Sunny Slope they will foul the Harbor.

Now about this famous flood gate, proposed to flush the Harbor. I have personally taken samples of the water at Hay Beach, and in the Harbor. There is a difference of as much as ten degrees Fahrenheit in the temperatures, and even in summer it is at least three degrees. The Harbor and the wetlands, by wetlands I mean lands which are flooded from time to time, are the breeding grounds of game birds, shore birds, fish, and numberless marine organisms vital to the food chain. These are very sensitive to temperature changes. Some fish will not breed if the temperature drops below seventy degrees even for an hour. All fish have an optimum temperature, and the Harbor evidently has it.

Next, salinity. I have tested that personally. The Bay is markedly more saline than the Harbor. Again, this is a breeding condition.

If we tamper with the temperature and salinity by attempting to flush the Harbor, you can kiss the harvest, or most of it, good-bye. Why not, therefore, prevent pollution, rather than try to clean it up after unwisely allowing it to take place?

Do you seriously think that the taxes on Sunny Slope and Hay Beach will pay for a piped water supply, a sewage treatment plant, a flood gate, the ruin of sport, the loss of the marine harvest, and the deterioration of Shelter Island into one more artsy-craftsy honky tonk?

It is my experience that when people understand, people act intelligently. The government of the Island has always tried to understand, and has, so far as I am aware, acted with forethought and intelligence. It is my pleasure to give you the benefit of this investigation without cost, in partial appreciation of the many happy years which all of our family has enjoyed on Shelter Island. Professors Kelland and Ramsdell have also extended this courtesy to you.

Most cordially yours,
Mary W. T. Arny
Associate Professor of Biology

Sample Letter to the Editor

4288 Wood Blvd.
Grangerville, Ohio 45679
May 18, 1971

The Editors, *Grangerville Times*
P.O. Box 4783
Grangerville, Ohio 45679

Gentlemen:

    As laudable as it is for the School Board of Grangerville to propose funding of future "ecology trips" by students in the biology courses at Grangerville High School, I must protest. It seems to me that students would learn more about ecology, develop a better sense of public-spirited cooperation, if they could see that same money being used for actual ecological improvement. For example, wouldn't it be better for town taxpayers' dollars to go toward the establishment of a glass recycling center than toward giving students bus trips to the shore refuge areas?

    Granted, the students will learn a good deal about their environment by taking field trips like those which are scheduled for the coming year. But wouldn't they and the town ultimately benefit more from a glass recycling center? They have such a center now in Hadleysburg, and the citizens there of all ages have been using it enthusiastically.

    Ideally, of course, the town could sponsor *both* the school trips and the glass recycling center; but as long as different goals compete for the same tax dollar, I would urge my fellow citizens of Grangerville to join with me and petition the town board to investigate the creation of a recycling center.

                                                              Sincerely yours,

                                                              S. Gebb Pettingill

COMMUNITY-ORIENTED WRITING PROJECTS

In addition to knowing how to write essays and letters, students ought to be looking for new ways in which they, as writers, can aid in the fight for a better environment. Maybe there are flyers to write announcing events of ecological importance—a company's laudable decision, for example, to add an emission control system to a cooling tower, or a university's decision to sell its stock in a company known to be irresponsible in its treatment of the environment—or of meetings of interest to ecology-minded citizens.

You might want to help write posters announcing the opening of ecology centers, recycling centers, trash clean-up drives, and so forth. There is endless work to be done. There are innumerable people to write to, not simply to accuse, but also to congratulate. Being ecologically responsible calls for being imaginative, and calls, too, for finding developments to praise and support as well as those to condemn and attack.

5
A GLOSSARY OF
ECOLOGICAL TERMS

Ecology, like every profession, has its jargon. To write persuasively within the field, you have to be familiar with its special language so that you can understand the issues before trying to communicate them to others.

This glossary is a compendium of biological terms which are ecologically oriented. It will help you to write about ecology with knowledge and assurance.

M. T. A.

A

accidental: A species which appears in a range where it does not ordinarily occur; for example, flamingos have appeared in the New Jersey Meadows.
acid bog: See *bog, peat bog.*
acid soil: A soil that tends toward acidity. Important because some species such as azalea will not grow or live elsewhere.
acre foot: Amount of water required to cover one acre to a depth of one foot.
actic: See *littoral.*
actino-: A prefix indicating radiation, particularly of light.
actium: An animal plant community peculiar to rocky marine shores.
activated sludge: Sewage disposal activated into decomposition by bacteria and protozoa.
adaptability: Quality of an organism which permits it to adjust.
adaptation: The act or change of structure which permits adaptability.
additive: Something added to improve function.
adsorption: Attachment to a surface; for example, impurities are adsorbed to a charcoal filter. (Commonly and incorrectly confused with absorption.)
adventicous: Species, organisms which have invaded from a great distance and established themselves. Uncommon.
adventitious: Same as adventicous; also refers to organs appearing in unusual places, such as the prop roots of corn which arise on the stem.
adventive: A plant out of its normal place, not native and remaining static in population.
aeolian: Pertaining to wind, such as aeolian deposition of soil. See *loess.*
aerobic: Needing oxygen to survive.
aeroplankton: Microorganisms borne upon air.
aestival: Also spelled *estival.* Having to do with summer.
age and area, rule of: The hypothesis that species which are widespread are older than species of limited range.
agro-: A prefix meaning having to do with soils, as agriculture.

aliquote: The temperature constant for specific portions of a life cycle.

allochthonous: Describes materials formed or created outside the environment in which they are found; opp. autochthonous.

alluvial: Materials consisting of clay, sand, or gravel deposited by flowing water.

altricial: Born dependent, for example, robins.

amorphous: Shapeless.

amphi-: A prefix, loosely, two kinds, as an amphibious plane can land on land or water.

amplitude: The range of tolerable conditions for an organism.

anaerobe: An organism that will not tolerate oxygen. Tetanus is an anaerobe; thus deep puncture wounds are dangerous.

anchor ice: Ice at the bottom of a stream. This occurs rarely since most ice floats (otherwise the oceans would be solid).

anoxia: Oxygen deficiency.

anthro-: A prefix indicating man.

aphotic: No light.

aqua-: A prefix indicating having to do with water.

aquifer: Water-bearing strata.

association: A plant or animal community composed of organisms frequently occurring together, such as spruce-fir associations.

asymptotic: Maximum size of a population.

autecology: The study of the relationships of individual plants to their environments.

autochthonous: Formed within the environment; see *allochthonous*.

available water: Water within the vadose zone that can be utilized by plants. Much of the water in this zone is attached to soil particles and is therefore unavailable.

B

barrier: Any of a variety of physical, chemical, or biological occurrences that prevents the migration of a species. Oceans, deserts, and mountains are examples of barriers.

biocycle: Life cycle.

bioecology: The study of the households of plants and animals and their interrelationships.

biological clock: The involuntary mechanism in living things which causes them to respond to time factors such as tides. Its cause is not understood. It is familiar to travelers on planes who can't adjust rapidly to new time zones.

biological control: As opposed to chemical control. Male insects being sterilized and released so that females do not reproduce. The introduction of species which eat other species and hence control them.

biological factor: The influence of biological interactions as opposed to physical, geographical, or chemical influences.

bio mass: The total quantity of living organisms in a specific volume of soil, air, or water. May also be a measure of the number of a certain species in the same.
biome: A specific type of environment, as the open meadowlands.
biosphere: The area between the upper atmosphere and the under soil in which life can and does normally exist.
bivalve: A sea creature with two shells which are hinged together, for example, clams and their kin.
bog: A swamp or marsh; acid bogs have acidic water.
border irrigation: Irrigation by means of flooding from border dikes, built of earth to hold water against need.
brackish: Slightly salty. Salt marshes and some places where streams and oceans meet are brackish.
breast height: 4.5 feet; 1.3 meters; d.b.h. (diameter breast high). The point 4.5 feet above the soil at which measurements of tree diameter are customarily made; also D.B.H.
browse: Low-growing twigs, buds, and young shoots of woody plants (line); food for deer, rabbits, and other animals.
buffer (species): Plants and animals that act as alternative food, thus reducing the demand on specific food species.
buffer (zone): A neutral place, separating influences, as the D.M.Z.

C

c_{14} **dating:** A means of determining the age of organic materials from an analysis of the remaining radioactive carbon found therein.
canopy: The covering created by treetop leaves.
capacity (adaptive): The elasticity of an organism which makes it possible to adapt to varying conditions.
capillarity (capacity): The tendency of liquids to move upward through small tubes or interstices.
carbon cycle: The motion of carbon from air or water to the plant through photosynthesis or the building of carbon compounds in plants or animals and its return to a source through disintegration or respiration.
carnivore: Meat eater.
carrying capacity: The greatest number of wildlife species which a given area can sustain in the most adverse natural conditions.
casual species: Species that appear unpredictably in a given area, as evening grosbeaks in the Middle Atlantic States.
catabolism (may also be spelled **katabolism**)**:** Metabolic breakdown usually involved in recombinations of elements, or in the direct production of energy.
catchment basin: Often seen along highways and usually fenced. They are dug to hold water which runs off the surface so that it can be absorbed by the soil instead of run off in sewers to rivers and the oceans.

cenospecies: The result of interbreeding species such as mule, tiglion, wolf dog, leopon, many of the avifauna. These forms are rarely fertile.

chaparral: Usually dense, low, scrubby plant life, as scrub oak and thorny shrubs. A specific biome found in the southwestern United States.

check dam: A small dam which promotes silting by slowing a stream. A beaver dam is a natural example.

chemotropism: The tendency of organisms to turn toward or away from certain chemicals, as plant roots will grow away from nails containing copper.

chinook: An Indian term for the warm, dry winds on the east slope of the Rockies. It is sometimes applied also to a warm, moist southwest wind along the coast of Washington and Oregon.

circle of vegetation: The sum total of species and communities of plants confined to a natural vegetative unit, for example, a nunatak where glaciation has not extended.

clay: Particles of mineral soil less than 0.002 millimeter in diameter. Also soil containing less than 45 percent sand, 40 percent silt, and 40 percent or more of clay particles.

clear cutting: Cutting down a forest lock, stock, and barrel.

climax: The most highly developed point a biological community can reach. As in the Middle Atlantic States the hardwood deciduous forest.

clone: Progeny produced by culture from a single cell or by parthenogenesis or asexual reproduction. Cuttings are clones.

closed community: A place where no available niche is left for invading organisms.

coliform bacteria: Bacteria common to the intestinal tracts of warm-blooded animals. Not in itself harmful, but an indicator of pollution, also called B. coli or E. coli.

colonial: Organisms that live in groups. They may be separate as nesting gulls, or together as coral polyps, but they are not dependent upon each other for survival.

commensal(ism): Living together to mutual benefit, or at least benefit to one organism without harm to another as barnacles on whales.

community: A group of organisms, not necessarily of the same species, living together.

competition: The struggle for survival arising between two individuals or species with the same needs.

conditioning: Teaching responses which result in adaptation (environmental); for example, putting young plants in progressively colder spots before setting them in the garden.

conservation: Saving things so that every organism including man may have a better life. Loosely used, protecting areas for public use.

consociation: A climax community which has a single dominant species.

consumer (organisms): Organisms that do not produce their own food. All animals as opposed to plants, which do make their own food.

contaminate: To make foul and unusable.
continuous grazing: Keeping animals on the same meadow or range without ever giving it a chance to recover. Good farmers shift their flocks and herds from meadow to meadow.
contour: An imaginary line connecting places of the same altitude. Often shown on maps.
control factor (limiting factor): The factor that creates the limit to which a process may go. If a candle is lighted and placed under a tumbler it will go out for lack of oxygen, hence oxygen is the limiting factor. Food, temperature, and water are other common limiting factors.
corridor: A land connection which enables species to migrate, for example, the former land bridge across the Bering Strait.
cover: A place to hide and to find shelter from the elements.
crepuscular: Active at twilight and dawn, as are bats and screech owls.
culled forest: As opposed to clear cutting, only those species or sizes desired are cut.
cycles (population, air, water, nitrogen): The phenomenon of anything going in circles, as nitrogen to fertilize grass, to feed cow, to dung and urine, to soil, to plant, *ad infinitum.*

D

D.B.H.: Diameter breast high (4.5 meters above ground).
D.D.T.: Dichloro-diphenyl-trichloroethane, which is lethal and spread all over the globe and must be banned world-wide.
desalinization: Removal of salt by artificial means.
desert pavement: Pebbles and rocks left after fine materials have been blown away or otherwise eroded.
deserta (xerophytes): Plants adapted to poor water availability such as cactus.
dewpoint: The point at which soil and air temperature differ sufficiently to precipitate the water out of the air in the form of dew.
diapause: Similar to hibernation or estivation. A period of lowered activity and metabolism which carries a creature through unfavorable climatic conditions.
diastrophism: The geologic process which deforms and dislocates the earth's crust, resulting in such formations as mountains and sea basins.
diatom (aceous earth): A plant form in either fresh or marine water usually microscopic having two outer shells which fit over each other as a box and its lid. Dead ones settle out and diatomaceous earth, also called fuller's earth, is formed.
differential species: A species that inhabits a special kind of locale, and thus identifies the locale by its very presence. Some plants are uranium indicators. Also called indicator species.
dimorphism: Having two or more forms, as the leaves of a sassafras tree.

dinoflagellate: One of a group of chiefly marine planktonic, motile algae capable of sudden population explosions. Blooms of certain species cause the "red tide."
disclimax: What is left of a true climax after man or domestic animals have altered it.
disoperation: A situation in which organisms harm one another because of an interaction; thus trees are stunted by browsing and over-browsing results in stunted or even starving deer.
dispersal: The spreading of organisms.
disphotic zone: That depth in water where there is not enough light for photosynthesis to take place.
dissemination: See *dispersal*.
diurnal: Organisms most active in the daytime; most birds are diurnal.
divergence: See *succession*. Divergence then is an increasing unlikeness of seres in a succession. Seres are series of stages in ecological succession.
dominance (ecological): The situation in which one species is overwhelmingly in possession of a biome or community.
dominance (social): The animal which is at the top of the flock, pack, or herd, is socially dominant. It is referred to as the alpha organism in terms of peck order.
dormancy: The condition of an organism being at rest, usually seasonal, as roses are dormant in winter in New York State.
drainage basin: That portion of a continent which constitutes a large natural drainage area of that continent such as the Hudson River drainage basin.
drought resistance: The ability of an organism to resist drought.
duff: The leaf litter and decomposed material on the forest floor.
dune: Large heaps of sand driven into this form usually by the wind.
dystrophic: A stage in succession of aquatic ecosystems in which high contents of humic materials and high oxygen consumption combine to cause a brown coloration of the water.

E

ecocline: Progressive changes in flora and/or fauna of an area resulting from changes in slope and exposure.
ecological amplitude: The extent of differing conditions under which an organism can live or a process function.
ecological bonitation: The number of a species estimated to be in an area. Beware of such estimates, which are often off as much as 50 percent.
ecological equilibrium: The balance of nature.
ecological equivalence: Animals or plants which require similar niches and therefore can displace each other; thus starlings have displaced bluebirds.
ecological factor: Anything in the environment that has an effect, good or bad, on the organisms living there.

A Glossary of Ecological Terms 59

ecological indicator: See *indicators*.
ecology: The study of "households" of living organisms together with their interrelationships, and the interrelationship with the environment.
ecospecies: A species made up of one or more interbreeding ecotypes; really a taxonomic species.
ecotone: Transitional area between two plant communities.
ecotype: A subspecies or race especially adapted to a particular set of environmental conditions, for example, many of the Galápagos species.
ecto-: a prefix meaning outside.
ectoparasite: A parasite which lives on the surface of a host, for example, a louse.
ectophagous: A species which eats from the outside such as a rabbit eating clover.
edaphic: Having to do with the soil.
effluent: Properly, the outflow from subterranean storage, now by common usage liquid, or liquid-carried contaminants.
encinal: A group of evergreen oaks.
endemic. Living only in a certain area, as the kangaroo is endemic to Australia.
endo-: A prefix meaning within.
energy flow: The taking in, conversion, use, and passage of energy through an organism or system.
energy transformers: Animals and plants which take in and pass on energy, as algae to fish to raccoon. In the vernacular, the flow of energy, usually through a food chain.
entropy: The downgrading of energy in a system; a measure of the disorder in the system's operation.
enzyme: A catalyst that will speed up or slow down chemical activity in the organism. Enzymes are usually specific; for example, diastase, found in the saliva, breaks down starch.
epibiotic: A species which is endemic and has survived from a former species.
epilimnion: The upper layer of water in a pond or lake.
epiphyte: A plant growing on a nonliving or living support but taking no food from it such as Spanish moss.
epizootic: Pertaining to any disease which is epidemic in animals, especially animals of the same species.
erosion: A geologic and/or meteorologic process whereby particles of soil or rock are detached from their base and carried away. Rainfall, wind, and flowing water are common agents of erosion.
erosion pavement: Stones or gravel on the surface, after finer material has eroded away. See also *desert pavement*.
escape: A plant found in the wild which is the progeny of a cultivated plant. A bird or animal which has escaped captivity.
established (also establishment): The condition in which a species moved to a new location grows with success.
estival (also spelled aestival): Having to do with summer.

estivation: Decreased metabolic activity in organisms during the summer; opposite season to hibernation.
estuary: That portion of a river in which fresh and salt water intermix in tidal zones.
etiolation: The condition of a plant grown with inadequate light in which the plant yellows from lack of chlorophyll and becomes abnormally elongated.
euphotic zone: The upper portion of a body of water in which there is adequate photosynthetic light
euplankton: Usually zooplankton. Animals in the plankton. (See also *plankton*.)
euryhaline: Describes organisms able to adapt to wide ranges of salinity.
euryphagous: Being able to select from a wide range of choices for food. Man is euryphagous; opp. stenophagous.
eutrophic: Describes lakes or ponds characterized by oxygen depletion, resulting in large accumulations of nondecomposed organic matter. This natural occurrence may be sped up by artificially introduced nutrients.
evolution: Genotypic changes in organisms resulting from selective natural breeding through the survival of the fittest.
evolution, convergent: The situation in which evolution produces similar adaptations to similar environmental conditions in widely diversified species.
exclusive species: A species which lives in association with one biome and seldom is found elsewhere.

F

factor: Anything that influences the environment or the organism.
facultative: The ability of an organism to live under several conditions, as facultative anaerobe; a bacterium which can live with or without oxygen.
fallout: Solid materials which are precipitated from the atmosphere, as radioactive fallout.
fallow: To lie fallow. The condition of a field not planted but weeded down for later planting, really a resting field.
fauna: All the animals inhabiting a given section in either time or space.
feedback: Return to original position after function is performed, as nitrogen used by plants and returned to soil through decay.
feral: A condition in which an organism has escaped cultivation and become established in the wild. Pigeons are feral birds.
field test: A test made in the natural conditions of the field, hence not artificially controlled.
filter: Any physical, chemical, or biological occurrence which retards migration of a species. Mountains, dams, rivers may act as filters. See *barrier*.

firebreak: A strip of land which has been cleared to prevent wildfire from spreading.

firn or névé: Compacted granular snow, usually at the head of a glacier.

fixation: The holding of a soluble substance in a relatively insoluble form, as nitrogen fixation by bacteria in the roots of many legumes.

flume: A device for carrying water across a stream or depression.

flyway: The path followed by migratory birds, as the Pacific flyway.

food chain: The progress of food from a primary producer to the ultimate consumer; a sort of biological farmer in the dell. A part of a cycle, pyramid, or web.

food web: The result of interlocking food chains forming a complex system of interdependent organisms.

forage: As a noun, wild food. As a verb, to seek available food; thus children forage in the icebox or cookie jar.

forest floor: Plant and animal residues on the ground under the trees.

formation: The set of characteristic natural climax and subclimax communities occupying an area as determined by climate and/or soil conditions.

fossorial: Burrow-digging, as by chipmunks. There are also some fossorial birds such as bank swallows and Leach's petrel.

freemartin: A sterile female, usually a calf and usually the twin of a male. Female imperfect sexually.

front: The point at which warm and cold air masses meet, usually visible due to cloud formations at the boundary.

fucoid: Resembling a seaweed, from fucus; the common algae with bladders that you popped as a child.

fumarole: A hole in the earth from which gases visibly erupt. Classic example, the Alaskan Valley of Ten Thousand Smokes.

G

gamete: A sex cell, as an egg.

gametophyte: The generation which produces sex cells, usually associated with clear-cut alternation of generations as found in ferns and mosses.

gastropod: A single-shelled mollusk, as a conch.

gene pool: Easily available genetic material as found in isolated areas or, speaking broadly, all genetic material.

gestation period: The period of time from conception to birth.

gramineous: In the family of the grasses, as wheat, bamboo.

grazing: Eating live grassy or herbaceous plants.

ground water: Water in the soil or rock strata.

growth form: The form taken on by an organism as an immediate response to environmental factors. See *ecotype*.

H

habit: The general growth pattern, as ivy has a climbing habit.
habitat: The type of environment inhabited by an organism.
halic: Saline soil and its associated organisms. The hal- prefix is applied to many other terms.
Holocene: The recent geological epoch or series.
holophyte: An organism whose primary source of energy is the sun.
holozoic: An animal which eats plants or other animals and has an internal digestive system.
homeostasis: The condition of an organism which acts according to its nature over a considerable gradient of difference. Man is such an organism.
homeothermic: An animal that maintains a reasonably stable body temperature under external temperature variations.
homologous: Deriving from the same origin, as arms of man and flippers of whale; opposed to analogous, having the same function but different origins, as wings of bird and wings of butterfly.
hook order: Peck order or social ladder of horned animals.
hormone: A secretion, usually of a special gland, which acts in portions of the body distant from it, for example, adrenaline, thyroxin.
host: A living creature which supplies benefits to an organism of another species; dogs are hosts to round worms.
humus: Soil having as its origin decomposed organic matter.
hydraulic equilibrium: That state in which equal pressures prevent soil water flow as in places where salt water intrudes into fresh water wells because of lack of hydraulic equilibrium.
hydro-: A prefix meaning water, as hydro (water)+tropism (turning toward). Most plant roots turn toward water and hence are hydrotropic.
hydrocole: An animal, fresh water or marine, which lives in water such as a fish or a whale.
hydrogen ion concentration: Usually written pH. A scale of acidity and alkalinity. Most familiar to the layman in testing his swimming pool.
hygric: Wetness of a habitat, or ecosystem.
hypertrophy: Unusual to abnormal enlargement of an organ or part, for example, elephantiasis is a form of hypertrophy.
hypo-: A prefix meaning under, as hypodermic, under the skin, and hypothetical, underlying a thesis.
hypolimnion: The water in lakes which does not circulate through most of the year because it is too deep down and therefore at a fairly constant temperature. It does circulate during spring and fall overturn.

I

idiobiology: Biology that confines itself to the study of individuals.

imago: Refers to mature insects; the butterfly, not the caterpillar, is the imago.
impoundment: A man-made body of water.
imprinting: A newly hatched chick first sees a cat rather than the hen. It follows the cat as its mother. The chick is said to be "imprinted."
indehiscent: Does not break open when ripe. (Botanical.)
index species: An organism which because of specialization lives in a highly specialized location; thus cattails always indicate that a marsh is fresh.
indicators: Organisms or species which are found naturally only under certain environmental conditions. The presence of those organisms indicates existence of those conditions. For example, oases are indicators of water.
industrial river: Euphemism for an open sewer!
infraneuston: An animal which is located on the underportion of the surface film of fresh water.
infusoria: Usually ciliated one-celled animals, such as the paramecium.
insect vector: An insect which transmits a disease without acquiring it.
instinct: Usually a reaction which is unlearned (but this is very much under dispute).
intertidal zone: The zone between high and low tide mark.
invertebrate: Lacking a spinal cord, as jellyfish.
in vitro: Cultured apart from the parent organism, as test tube babies; opposite of *in vivo*, regular babies.
ion-exchange: The replacement of one kind of ion by another, an ion being a portion of a chemical compound which is electrically charged, as water H_2O written HOH. There is then the hydrogen (H) ion and the (OH) or hydroxyl ion.
iso-: A prefix meaning same or equal.
isobar: A line drawn on charts to indicate points of the same barometric pressure. Also isotherm, for equal temperature, as on weather maps.
itograph: An instrument which counts the number of trips birds take to and from the nest. Sometimes also indicates time and direction.

J

Jordan's rule: Cold-water fish have more vertebrae than warm-water fish (a broad generalization).

K

K: Symbol for carrying capacity.
kame: A hill or mound of materials deposited by meltwaters from a glacier.
kar herbage: The tall plants that are found in fertile hollows at high altitudes.

karst: Limestone areas characterized by sinkholes and solution depressions.
karyokinesis: Mitosis.
karyotype: The visual aspect of chromosomes other than those found in the gametes.
key-industry: Describes herbivores which are the staple of a food chain.
Krebs cycle: A long and involved interlocking series of aerobic respiratory reactions.
krummholz: Eroded scrubby trees and shrubs usually at high elevations. The result of constant pressure by wind-borne particles of ice or sand.

L

lacustrine: Produced in, living in, or pertaining to lakes.
lagg: A sphagnum or peat bog with a center higher than the edges.
land bridge: A connection between large land masses used for migration, as the Aleutian Islands are the remnants of a land bridge.
lasion: See *periphyton*.
laurilignosa: Subtropical rain forests where laurel is likely to predominate.
layer: Applies to both plants and soils—in plants a group of the same height, as grass, shrubs, trees.
leaching: The removal of soluble materials from the soil by water action.
leaf mold: Decayed leaves, usually finely divided.
lentic: Bodies of water, such as ponds, which contain standing water.
lenticel: An aperture on a woody stem used by the plant for gaseous exchange.
lethal gene: A gene which inevitably causes death.
liana: Vines such as Tarzan swings on.
lichen: A plant, usually small and greyish green, which is actually a combination of an alga and a fungus. They are very important to pollution studies, being highly sensitive to gasoline fumes and sulfur dioxide particularly.
Liebig's law of the minimum: Whatever necessary factor is in smallest supply determines the growth and reproduction of an organism.
life zone: A belt with distinctive flora or fauna. May be determined by either latitude or altitude, as a nunatak.
light quality: The composition of light by wave lengths.
lignin: The substance which makes cells woody and hard.
limiting factor: Factor which dictates the possible growth of a plant, animal or community. Water is the limiting factor in the desert.
limnology: The study of fresh-water ecosystems.
Lincoln index: Population size as estimated by a count of marked animals, often of debatable value.
line-intercept technique: An arbitrary line is drawn across an area and the species on the line are tabulated. One technique for estimating plant populations.

line-plot survey: Similar to the line-intercept technique but plots are squared off at random along the line, and plants within the plots are tabulated.
line-transect: Same as line-intercept.
lister: A specialized plough which throws the soil so that water-holding furrows are formed.
list quadrat: A rectangular sample area. All too often miscalled "quadrant."
litho-: A prefix meaning stone. A lithophyte is a plant that grows on stone, for example, a lichen. But not all lichens grow on stone.
lithosphere: The solid portion of the earth's crust between air and water and the earth's core.
littoral: In fresh water, the area from the shore to the last point at which rooted plants are found. The marine littoral is from the high water mark to the boundary of wave action and light penetration, but more commonly refers to the area directly covered and uncovered by the tides.
liverwort: A group of plants between the algae and the mosses, such as Marchantia, important as bridging the water-to-land evolution.
llorano: Fog caused by "northers" along the Gulf of California in winter.
loam: A sand-silt-clay soil combination, but less than half sand.
loess: Fine soil deposited by wind.
loss-on-ignition: The weight loss caused by burning out organic material from previously thoroughly dried soil.
lotic: Running water—hence the generic name of Lotor the raccoon.
luminescence: Light caused by biochemical action rather than heat.
lunar periodicity: The correlation of biological activity and lunar (moon) periods.
lycopod: A plant group, the most familiar of which is ground pine. This is often used as roping at Christmas, and is usually a protected species.
lysimeter: An instrument for measuring leaching. See *leaching*.

M

macro-: A prefix meaning large.
macronutrient: A substance used for food in relatively large quantities, for example, rice in Japan, calcium for clams, and grass for cattle.
malacology: The study of mollusks; a subdivision is conchology, the study of shells.
malthusian: The concept proposed by Thomas Malthus that organisms increase in geometric progression while food increases in arithmetic progression.
mammal: An animal which has hair and suckles its young. You're one.
marsh: A swampy area in which rushes, sedges, grasses, and cattail are the dominants.

marsupial: An animal which brings forth live young that are then suckled in a pouch. The opossum is the only North American marsupial.
mast: Nuts. It was the loss of the beech mast more than hunting which caused the passenger pigeon to become extinct.
megatherm: An organism which requires continual high temperature to survive, for example, bananas.
meiosis: The process of cell division by which the chromosome number is halved in the production of egg and sperm. Sometimes called reduction division. The half number of chromosomes is called the haploid number.
melanin: A dark pigment conspicuous on the wing tips of long flight birds; it strengthens the feathers.
melanism: A dark form of a normally lighter animal as the black grey squirrels common in Princeton, New Jersey. Caused by a deposit of melanin.
melliphagous: Feeding on honey, for example, Winnie the Pooh.
Mendel's law: A complex set of mathematical proportions which hold in the offspring of organisms with definitive characteristics.
meristem: Plant tissue which divides to form new cells.
meso-: A prefix meaning moderate.
mesophyte: A plant which grows in moderate climatic conditions.
mesosaprobic: Aquatic environment where decomposing vegetation reduces the oxygen content drastically.
metamorphosis: The change of a creature from one form to another, for example, caterpillar to butterfly.
meteorology: The study of weather.
microhabitat: A small habitat, say a tree stump.
micronutrient: Trace element. Only needed in small quantities yet vital; thus lack of boron in many plants causes necrosis of the meristems.
microphagus: Large animals which eat ridiculously small things for their size—elephants eating peanuts.
microtome: An instrument for cutting very thin sections for microscopic study. A kind of glorified ham slicer.
migrant: Something that moves around a lot.
mille-: A prefix referring to a thousandth part.
mimicry: The characteristic behavior of looking like some other organism, thus escaping predation. A leaf hopper that looks like a rose thorn is practicing mimicry.
minimum: See *Leibig's law*.
mitosis: Cell division of all cells except those producing egg and sperm.
moisture tension: The force at which soil water is held.
monocotyledon ("monocot"): A plant whose seed is in one section as is corn's, as opposed to lima beans, whose seeds are clearly in two sections. Monocots are seldom woody; palms and bamboo are exceptions.
monophagous: Lives on only one or a very few things. Important in food chains; for example, the Everglade Kite eats a certain snail almost exclusively; the snail is vanishing; ergo so is the Kite.

montane: Located in high mountain regions.
morphology: The study of the form development and structure of creatures; includes plants.
mulch: Plant residue layers. Sometimes extended to other materials. Used in gardens to cut down weeds and retain moisture.
multiparous: Animals which as a rule bear more than one young at a time.
multivoltine: Having more than one litter per season. Gerbils are multivoltine.
muskeg: Northern North American sphagnum bog. Sometimes a few shrubs and low trees, likely to be spruce or cedar.
mutation: An inheritable departure from the norm.
mycorrhiza: Fungi which are associated with specific plants and essential to their growth. Pink lady slippers are very difficult to transplant because their specific mycorrhiza are absent in most soils.
myrme-: A prefix meaning dealing with ants.

N

nanism: Abnormally small appearance of plants due to extreme conditions.
nanoplankton: Small plants or animals which will pass through number 20 silk bolting material.
nastic motion: A plant response due to diffuse stimulus and probably hormonal in origin, as the closing of sensitive plant.
natatorial: Having to do with swimming.
natural selection: The survival of the fittest.
neap tides: Lowest tides. They occur at the first and last lunar quarter.
necrosis, necrotic: Having to do with isolated dead portions of an organism.
nectar: The sweet liquid produced by flowers. Not to be confused with honey, which is processed by bees.
necton: Motile organisms found near surface of a body of water.
nekton: Fish, crabs, and other sea dwellers which are strong swimmers.
neoplasm: New protoplasm of abnormal form.
nepho-: A prefix indicating having to do with clouds; a nephometer measures overcast and a nephoscope measures speed of cloud movement.
nested quadrats: Quadrats, like nested boxes, are getting bigger and bigger. Used to determine ideal size for a study, or to study various forms of vegetation with largest quadrat for tree samples and smallest for herbaceous plants.
neuston: Organisms depending on surface tension of water for their support, for example, mosquito larvae, water striders.
neutralism: The presence of varied organisms which do not affect each other. A dubious term.

névé: Granular packed snow, usually refers to glaciers, but sometimes used to describe trails.

niche: A small concavity used to contain statuary or *objets d'art* but carried into biology to indicate a specific opening in the habitat occupied by an organism. Functional niches and locational niches should be differentiated.

nid-: A prefix relating to nests.

nidicolous: A most unusual word. Really means altricial. Refers to birds which hatch naked and helpless, such as robins; opposed to precocial, hatched ready to go, as chicks.

nitrogen cycle: The path nitrogen follows: from soil and air to plant, to cow, to manure, to soil.

nitrogen fixation: The process by which highly specialized bacteria make atmospheric nitrogen available to plants.

nivation: Erosion caused by snow.

nocturnal: Active at night.

nomina conservanda: Almost archaic but means names used by common consent, though they do not follow regular rules of nomenclature.

nuciferous: Bearing nuts.

nunatak: A portion of the earth which, although in an area that was glaciated, escaped glaciation. There is a nunatak on Katahdin. Nunataks are of interest to botanists because they have a very specialized flora.

nutation: The spiral reaching motion of plants looking for something to cling to. Also called nyctitropism.

nymph: A stage of insect development between egg and adult.

O

obligate parasite: A parasite which has to have its host to survive.

oecology: The study of the household. (The preferred form is ecology.)

oestrus: Period of heat in females of the mammalian group.

omnivorous: Not restricted to either animals or plants for food. Men and bears are omnivorous.

ontogeny: The development of the individual; opposed to phylogeny, the development of the race. Ontogeny recapitulates phylogeny; or the development of the individual from conception to adult form repeats the development of the phylum to which it belongs.

open community: A place where plants are thinly scattered and therefore there is room for other species to invade. Also a community with unfilled niches.

organelle: A specialized portion of a cell, such as a mitochondrion.

ornithology: The study of birds.

ortet: The parent organism from which a clone was derived.

osmosis: The motion, through a differentially permeable membrane, of materials of differing concentration. Motion takes place in both

70   Ecology: A Writer's Handbook

directions but the greatest motion is from the place of greatest concentration of the *solvent* to the place of least concentration of the *solvent*.

osmotic pressure: The pressure potential of osmosis.

oviparous: Laying eggs which are externally fertilized, as do frogs and fish, as contrasted to ovoviparous, where internal fertilization takes place, as in mankind.

P

paleobotany: The study of fossil plants.
paleozoology: The study of fossil animals.
paludal: Having to do with marshes.
palynology: The study of pollen.
pandemic: Something widespread; the common cold is a pandemic disease.
pathogen: A virus or bacterium which causes disease—in the vernacular, a germ.
peat bog: A bog in which the prevailing plant residue is peat which makes it acid. Peat is old sphagnum.
peck (pecking) order. The order of dominance. In most families, father, mother, oldest son, youngest daughter, dog, cat. In the army, the commander-in-chief down the ranks to the raw recruit. Exists in all social structures. Position in order, indicated by Greek letters with dominant individual called alpha.
pedology: The science dealing with soils.
pedon: A community which exists on a lake bottom.
pelagic: Species found beyond the three-mile limit in the open sea; the albatross is a pelagic bird.
Pennsylvanian: This one you could stumble on. It is not an inhabitant of Philadelphia, it is a geological era which started roughly 255 million years ago and lasted for about 25 million years.
perched water: The upper portion of free ground water, usually separated from the underlying ground water by an unsaturated layer of soil, marl, rock, and so forth.
perennial: A plant that lives three years or more. (Also hardy perennial, the same type of plant growing out of doors.)
periphyton: All the organisms attached to surfaces above the bottom of marine or fresh bodies of water, such as sea anemones on piles.
permafrost: Ground which is permanently frozen.
permeant: An animal that travels widely, for example, a robin.
pessimum: Opposite of optimum.
pesticide: A substance designed to kill creatures which man dislikes.
pH: Hydrogen ion concentration, a measure of acidity.
phag-: A prefix meaning eat.
phenotype: The characteristics of an organism as affected by its environment and its genes; loosely, what it looks like.

photo-: A prefix indicating having to do with light.
phylum: A major group used in classification; mammals constitute a phylum.
phyto-: A prefix meaning plant, as phytoplankton.
pioneer (organism): The first inhabitants of a naked area.
plankton: Weakly swimming or floating organisms, animal or plant.
plantigrade: An animal that walks on the whole foot, for example, man.
pluv-: A prefix meaning having to do with rain.
poikilotherm: An animal such as a snake which has no built-in temperature control.
pollution: Loosely, contamination of a habitat; precisely, the state of being soiled, fouled, or defiled. As opposed to contamination, the physical contact that causes pollution.
polymorphic: An animal having several distinct forms, as black and yellow swallowtails.
pothole: A hole formed in rock by grinding of another rock, usually in a stream.
precocial: Opposite of altricial. Born able to cope, as are ducks.
predator: An animal that kills other animals to live, for example, snakes, men, hawks. It is well established and acknowledged that man is the primary predator.
prey: That which is eaten by a predator. By the way, it is *preying* mantis.
producer: An organism which converts radiant energy to usable food.
profundal (area): Part of a body of water where no light penetrates. Aha, I have always suspected profound people!
protective coloration: Camouflage.
protozoan: A one-celled animal such as an amoeba.
psammolittoral: The sandy shore of a lake.
public domain: Land whose title is held by the federal government. Used loosely to mean no consent is needed; thus pictures of people on the street may be published without consent because they are "in the public domain."
pyramid of numbers: A diagram illustrating that animals on lower levels of a food chain are more abundant than those on successively higher levels.

Q

Q_{10} **rule (Van't Hoff rule):** The rate of response of an organism is often doubled or more for each 10°C increase in temperature (within limits). A vulgar application would be building a fire under a recalcitrant mule.
quadrat: A sampling area usually 10 meters square.
quantum: A unit of energy emitted by a particle in a beam of radiant energy.

R

race: A group that differs superficially from another group of the same species. (See also *species*.)

rad: A measure of ionizing radiation.

radio carbon dating: Determination of age by measurement of the remaining amount of radioactive carbon. Its accuracy is now seriously challenged and may be off as much as 700 years. This has been demonstrated by tree rung studies.

radiosonde: A free balloon carrying instruments and transmitting data.

ramet: An individual member of a clone.

random sample: A sample taken by pure chance, as by closing your eyes and sticking a pin in the map.

range: Area in which a species may be found.

raptores: Birds with curved talons for taking prey, for example, owls.

rasorial: Animals such as hens that usually scratch the ground for food.

red tide: A sudden bloom of certain dinoflagellates which makes water look red. It usually kills fish and was probably the phenomenon described as the water turning to blood in Exodus 7.

reed swamp: A plant community in shallow water, usually reeds, rushes, and cattails, some of which may extend onto dry land.

reef: Rocks or coral rock lying close to the surface of the water and often exposed at low tide.

reflex: An uncontrolled response to stimulus, as eye blinking when an object approaches the eye.

regeneration: The process by which an organism replaces lost parts or heals wounds, as some lizards can grow new tail tips.

resistance: The ability of an organism to withstand a condition such as drought, frost, or disease.

rheo-: A prefix indicating having to do with running water.

rhizobium: A bacterium, most commonly associated with legumes, which is able to make free nitrogen into compounds that can be used by the host.

rhizoid: The evolutionary forerunner of the root.

rhizome: An underground stem, as iris rhizomes.

rime: Frost which comes from moisture deposited by fog or clouds.

riparian: Land bounding water.

runoff: The portion of rain or melted ice and snow not absorbed by soil.

S

sabulicole: An insect that lives in sand.

salinity: Degree of saltiness.

salt marsh: A marsh where water is brackish or saline.

saltigrade: An animal with legs adapted to jumping, for example, a kangaroo.
saprophyte: A plant which lives on decayed organic matter and hence does not manufacture its own food, for example a mushroom.
savannah: Grassland with a few scattered trees and shrubs.
scat: Animal droppings, hence "scatological language."
scavenger: A creature which eats carrion and other refuse.
scion: A portion of one plant grafted to the root, stem, or branch of another.
sclerosis: Hardening by increase of lignin or collagen in plants and animals, respectively.
seaweed, marine algae: Do not confuse with fresh-water plants, many of which are highly developed and flowering. There are a *few* flowering marine plants.
sedentary: Tending to sit still, as opposed to sessile, which means attached to substrate.
sedimentary rock: Rock formed from sediment.
seepage: Water which passes through the soil along a surface, and the process by which it moves; thus, water *seeps* from the soil down a rock face.
semipermeable: Obsolete. Differentially permeable is now proper usage. Something which selects what can pass through it; for example, cell walls are differentially permeable.
serology: The study of serums.
sessile: Attached to the substrate.
settling basin: A basin wider and deeper than the rest of a stream, usually artificial, in which sediment is deposited.
shingle: Pebbles worn round by the sea, found at the seashore.
siblings: Offspring of the same parents.
sinkhole: A hole where surface water drains into an underground channel, commonly in limestone areas.
slime mold: A naked mass of fluid protoplasm of the phylum Myxomycophyta, acellular but containing many nuclei.
slough: Similar to a swamp but with deep mud.
sludge: Unpleasant muddy, oozy sediment.
smog: A combination of smoke and aerial moisture polluted in addition with noxious compounds.
snow line: The lower boundary of eternal snow.
softwood: A misnomer for coniferous wood.
soma: The aggregate of body cells excluding egg and sperm.
spat: The larvae of bivalves, as oyster spat.
spawn: Eggs of fish, some mollusks, and amphibians; also the mycelia of fungi used in propagation.
spay: Remove ovaries.
species: The group of the same genus which are fertile to each other for *more than one generation*.
speleology: The study of caves.
spelunker: One who explores caves.

sphagnum bog: An acid bog in which sphagnum predominates. Loosely, a peat bog.

spore: The asexual reproductive structure, as of ferns and mosses; opposed to the sexual seed.

sporophyte: The portion of a spore-producing plant that produces the spores.

stand: A group of plants, usually trees, of predominantly one type and/or age class.

station: A particular point at which a plant species is found, usually used in reference to rareties.

stilling basin: An excavation at the base of a fall or rapids which slows the current.

strain: A subdivision of a species, usually familial—as the Hapsburg strain.

stratification: A phenomenon which divides an area into separate layers. Lake stratification occurs due to temperature and/or density differences; forest stratification includes canopy, understory, shrub layer, herbs, roots, soil.

stumpage: Uncut timber, usually used to mean its standing monetary value.

succession: The orderly progression of associations from bare substrate to climax vegetation.

sweepstakes bridge: Accidental transport of a species, as rats to the Galápagos.

symbiont: An organism which lives in close association with another organism, sometimes to mutual benefit, sometimes to benefit of only one but no harm to the other, sometimes harmful to one and beneficial to the other, as in parasitism.

synecology: The study of the ecology of communities.

synergism: An exaggeration of effect when chemicals or drugs are used simultaneously; thus alcohol and barbituates even in minor dose may prove lethal in combination.

T

table land: A flat area raised from a plain and surrounded by cliffs or steep slopes.

taiga: Swampy forest adjacent to the tundra, usually coniferous.

tailings: Mine refuse.

taproot: A single large main root such as a carrot.

taxis: Motile response of an organism to external stimuli. Also a combining form, as phototaxis, chemotaxis, rheotaxis.

teleological: An attitude implying that something is shaped by its purpose. Fish are thought to be streamlined *so that* they can swim fast, when in fact fish can swim fast because they are streamlined.

temperature inversion: A situation in which ascending air heats instead of cooling. Effectively it holds smog down.

teratological: Relating to monstrosities, for example, thalidomide causes teratological effects.
territoriality: A behavioral phenomenon in which one organism "says," in effect, "stay out of my yard."
thermo-: A prefix meaning having to do with temperature.
tick: A member of the spider family which burrows into the skin, particularly in hairy areas. It is extremely dangerous because it carries Rocky Mountain spotted fever. It should be covered thickly with petrolatum (vaseline) until it can be removed by a physician. Do *not* try to get it out unless you are experienced.
trace element: A substance needed by organisms in minute amounts; boron, iron, and magnesium are some trace elements needed by man.
transect: A long, narrow area studied for analysis of life forms in a given biome.
transpiration: Giving off of water.
tsunami: Often miscalled a tidal wave but really caused by submarine earthquakes.
tuber: An underground stem such as a potato used by the plant for storage. Tubers may be used for vegetative reproduction.
tularemia: A rodent and rabbit disease easily transmitted to man through contact on an open wound. Do not *ever* touch a slow or dead rabbit, or in fact any slow or sick wild animal.
tundra: Arctic areas characterized by low-growing forms as shrubs and lichens and often underlaid by permafrost.
turnover: The shifting of layers of water and their constituent nutrients due to great temperature changes. Spring and fall turnover in reservoirs often causes drinking water to taste foul and appear turbid, a temporary and innocuous occurrence.

U

undergrowth: Those plants growing up between trees on the forest floor.
understory: Those trees of secondary height growing below the upper canopy, for example, dogwood.
ungulate: Not commonly used; a member of a taxinomical group of hooved mammals.
unicellular: An organism such as the amoeba composed of only one cell. The preferred term today is "acellular."
uniparous: Creatures producing only one egg at a time or bearing only one offspring; man is normally uniparous.
univoltine: A creature that has only one generation in a year. Dogs are univoltine, whereas mice are multivoltine.

V

vagility: Ability to disperse.
variegated: Patched or striped, usually with white, as ribbon grass.
vascular: Having veins, vessels, or other conduits which carry water or solutions.
vector: An organism that transmits a disease without itself acquiring it; for example, fleas are the vectors of bubonic plague.
vegetative reproduction: The natural propagation of a plant without seed as by the Bryophyllum.
vernal: Having to do with spring.
vernilization: Artificial lowering or raising of temperature, usually to cause germination in seeds.
virus: A structure that can reproduce only within the cell of a host. Whether it is a living or nonliving thing is under dispute. Rubella (German measles) is caused by such a structure.
viviparous: A creature which bears its young in a form not basically differing from that of the adult; for example, kittens are obviously immature cats.

W

watershed: The area from which water in a lake, stream, or pond comes.
water table: The top of the zone of saturation. That is, at and below water table level. All spaces in rock and/or soil are filled with water.
weed: Any plant that grows where you don't want it.
windbreak: Trees or other plants placed to protect homes or crops from the wind. Often these act as sound barriers as well.

X

xero-; xer-: A prefix usually implying a dry habitat; cactus is a xerophyte.
xylem: Woody tissue; for example, boards are made from the xylem of a tree.

Y

y chromosomes: Genetic units determining sex, found in humans only in the male.

Z

zone: A division of vegetation clearly delineated; also one of the climatic belts of earth. Zones given on the back of seed packets indicate planting times.
zoo-: A prefix indicating a relevance to animals.
zygote: A fertilized egg, plant or animal.

6
ESSENTIAL USAGE

When *Webster's Third International Dictionary* was published, an article reviewing it was entitled "Raviolis are in, Spaghettis ain't." The title called attention to the latest *usage,* which indicated that one could now pluralize ravioli but not spaghetti and that the word *ain't* was now becoming common (as dialect) in some places and was included, for the first time, in the dictionary. The title of the article conveyed the important fact that language, and thus our use of language, is always in a state of change. New words like *pizzaburger* and *split-level* are being added to our vocabulary constantly. The changes in a language are demanding; we need to know more. The point of including in this book a long glossary of ecological terms is to help people using what are, for them, new words. Although language changes, there are some forms of usage that are still generally considered better or preferable. This chapter, like the glossary of ecological terms, is meant to be a small book in itself and should be used as a dictionary or handbook of essential usage. The following list does not pretend to be exhaustive. Students unable to find the usage of a word or phrase should consult the *Oxford English Dictionary,* Fowler's *Modern English Usage,* or Miss Nicholson's Americanization of Fowler, the *Dictionary of American-English Usage.* All of these works and the usage glossaries of a variety of other textbooks have been consulted in compiling the list that follows. It is hoped that this list will be a practical guide to good writing.

accept, except: These two verbs are needlessly confused. The former means to receive and the latter to exclude. *Correct*: He *accepted* the suggestion politely. He *excepted* the fourth original clause from the final contract.
account of, account for: When account is used as a noun, it is followed by *of*; when it is used as a verb, it is followed by *for. Correct*: He made an *account of* the most recent catastrophe. His frequent absences from classes *accounted for* his failure. Both expressions mean to explain something.
advertise, advertize: The first spelling is preferred. (The same is true of *advertisement*.)
affect, effect: To *affect* is to act on, to produce a change in or an effect in. *Correct*: This accident may *affect* his eyesight for many years. To *effect* is to cause, to bring about. *Correct*: The transition to air transportation was *effected* last winter.
afterward, afterwards: Either spelling is acceptable. (*Afterward* is preferred in American usage.)
all right: This is the only acceptable spelling.
alot: No such word. The word *lot* means a group of associated things or people, and thus we can have a *lot* (two words) of them.
already, all ready: *Already* is an adverb. *Correct*: Peter has *already* left for school. The two words *all ready* are used to describe the condi-

tion of complete readiness. *Correct*: Is everything *all ready* for the Ecology Action meeting?

also: Never use to replace *and*. *Correct*: We had bananas, apples, *and* grapefruit. (*Incorrect*: We had bananas, apples, *also* grapefruit.)

altogether: This is the correct spelling of the adverb meaning completely. *Correct*: It was *altogether* ridiculous for the company to pretend it had any concern for the environment. The two words *all* and *together* are used to describe the physical condition of being in a group. *Correct*: The boys are *all together* in the truck.

among: Use when discussing more than two things. *Correct*: He was *among* the five best astronomers in the nation. *Correct*: Who *among* you, class, has solved this problem? Use *between* for two things only.

analyze, analyse: The first spelling is standard in America.

and/or: Considered awkward except in legal writing.

anti-: A hyphen is preferred when joining to nouns beginning with a vowel.

approve, approve of: When used in the sense of endorse or ratify, no *of* is necessary. *Correct*: The board *approved* the contract. When used in the sense of bestowing praise, *of* is needed. *Correct*: He *approves of* the way she dresses.

as well as: See *both*.

averse, aversion: Better to follow with the preposition *to*, not *from*. *Correct*: He has an *aversion to* litter-bugs.

awhile: This is the correct spelling of the adverb. *Correct*: I will operate the machine *awhile*. With *for* it becomes a three-word prepositional phrase, *for a while*. *Correct*: Why don't you *wait for a while*?

backward, backwards: Either is acceptable. (*Backward* is the preferred American usage.)

beside, besides: Use *beside* to mean alongside of. *Correct*: The ecology book is *beside* the lamp. Use *besides* adverbially to mean in addition (to). *Correct*: She does not really love him enough; *besides*, she wants to go to graduate school before she marries.

between: Never repeat in a sentence. *Correct*: *Between* the manager and the star, the show was a failure. (*Incorrect*: *Between* the manager and *between* the star, the show was a failure.) *Between* may lead directly to a plural. *Correct*: *Between* the teams. *Between* may also lead to two phases joined by *and*. *Correct*: *Between* the fox *and* the lion.

both: Follow with *and*, not with *as well as*. *Correct*: *Both* Fred *and* John were there. When using *as well as*, simply eliminate *both*. *Correct*: He is a successful composer *as well as* a talented pianist. Redundancy results when *both* is joined to phrases such as *alike, at once*, and *between. Correct*: It is *both* dangerous and exciting. (It would be redundant to write: It is *at once both* dangerous and exciting.) If *both* is used before a preposition, the preposition should be repeated; if the preposition is placed before *both*, it should not be repeated. *Correct*: *Both in* this country and *in Russia. . . . In both* this country and Russia. . . . It is also better to repeat the article before each of two things joined by *both*. *Correct*: *Both the* radio and *the* television are broken.

brilliance, brilliancy: They can be used interchangeably; however, the former is usually used to mean talent, genius, and so forth, whereas the latter is used when discussing light and colors.

but: Use *all but he,* not *all but him*. *Correct*: Thereupon *all but he* left the room. Never use *not* after *but*. *(Incorrect*: Who knows *but* that the country may *not* be here in ten years?) *Correct*: Who knows whether the country will be here in ten years? Note also that it is always redundant to write *but . . . however.*

by: Repeat as a preposition in front of a series of objects. *Correct*: The river was being ruined by chemicals, by garbage, and by selfish fishermen. Sometimes, when writing a longer series, the *by* need precede only the first and the last items: *Correct*: It was decided by the school board, the townspeople, the wardens of the church, the ladies clubs, *and by* the mayor that the town needed a glass recycling center.

can't help but: Omit the ungrammatical *but* and use a gerund. *Correct*: He *cannot help winning* every game. Sometimes it sounds better to rephrase the statement entirely: He is so good that he wins every game.

case: Many expressions involving the word *case* (usually because of colloquial habit) can work more efficiently without it. *Avoid*: In many *cases* the players signed contracts. *Use*: Many players signed contracts. Similarly, *avoid* expressions such as "in no case" (use "never"), "this is so often the case" (use "often"), "in many cases" (use "often").

cause: *Avoid* following with *is due to*; just use *is* (see *reason*).

choice: Never use as an adjective.

common sense: always two words.

concept of: Often superfluous. *Write*: pollution. *Avoid*: concept of pollution.

consider: Eliminate the use of either *to be* or *as* when using *consider* in the sense of to judge or to evaluate. *Incorrect*: He was *considered to be* the most brilliant man in his field. *Correct*: Jefferson *considered* democracy *to be* the natural way of man.

considering: It is acceptable usage to mean with reference to. *Correct*: *Considering* his handicap, he plays well.

contact: Avoid use as a verb. The correct expression is *to make contact with*. *Correct*: He *made contact with* Charlie in Manila. (Note, however, that *contact* can sometimes be used as a verb in a technical, primarily electrical sense: A spark appeared where the wire *contacted* the pole.)

continual, continuous: The first is used to express an action which "goes on" with interruptions or lapses of time. *Correct*: He continually fails to brush his teeth after every meal. *Continuous* denotes something that "goes on" *without* interruptions. *Correct*: There was a *continuous* hum in the basement because of the noisy hot water heater. (Think, too, of the sign in front of a movie theater reading *"Continuous* showings, noon to midnight.")

contrast: Something *contrasts* or *makes a contrast* (the first is preferable) *with* (not *to*) something else. *Correct*: The darkness of the mountain *contrasts with* the pale mist in the valley. Also note that whereas *contrast* concerns a difference, *compare* concerns likeness. *Correct*: This book *compares* favorably *to* (or *with*) his last one.

cooperation: No need to use the hyphen (*co-operation*). The same is true with similar words—*coeducation*, *coauthor*. If a hyphen will make a word more immediately understandable to the reader, then it can be used; for example, *co-worker* registers with the reader's eye more quickly than *coworker*.

counter-: It is never necessary to use a hyphen, but some authorities suggest doing so when the *counter* is being prefixed to a long word, for example, *counter-organization*.

criterion, criteria: The first is singular, the second plural. *Correct*: This *criterion*; these *criteria*.

data: plural only. *Correct*: The *data represent* the efforts of two years. Also, always write "many data," not "much data."

defense, defensible: Correct spelling.

desert, dessert: A *desert* is flat land in Arizona. A *desert* is an act or quality deserving reward or punishment. To *desert* is to abandon; *deserted* is abandoned. A *dessert* may be cake or pie. *Desert,* an adjective meaning uninhabitable, pertaining to the desert, is rarely used.

dullness: Always two *l*'s.

each: Each individual person or thing, thus always singular. *Correct*: *Each* of the boys on the team *has his* own locker. For more than one, use *all*. *Correct*: *All* the boys *have their* own lockers.

effect: See *affect*.

e.g., i.e.: The abbreviation *e.g.* (Latin, *exempla gratia*) means for instance. The abbreviation *i.e.* (Latin, *id est*) means that is to say. Therefore *i.e.* should never be used to introduce an example, and *e.g.* should never be used simply before rephrasing or clarifying. No comma is necessary following either *i.e.* or *e.g.*, but it is acceptable to use one.

egoism, egotism: The former is a more technical philosophical term; the latter is generally used to denote excessive self-concern.

eligible: Correct as spelled.

equally as: Omit *equally*. *Correct*: The first novel is *as* important as the second. Note also that it is *permissible* to say: It is *equally* good. (*Incorrect*: It is *equally* as good.)

etc.: Do not use *etc.* in the middle of a sentence. An *etc.* is best used at the end of a sentence in which the writer's further examples, were he to write them out, would be obvious in advance. *Correct*:

He did all the household chores—the ironing, dishwashing, sweeping, *etc.*

ever-: Use the hyphen. *Correct*: *ever-ready, ever-growing.* In general, however, it is better to find another way of writing the same idea, often making use of the word *always.*

ex-: Try using *former* instead (the *former* senator from Vermont).

except: See *accept.*

female, feminine: The first denotes that sex for any and all creatures, whereas the latter denotes womanliness.

firstly: No longer considered good usage (though *secondly, thirdly,* and so on, have survived).

focus: Plural is either *focuses* or *foci.*

forever: One word.

fulfill, fulfilled: This is correct spelling.

hardly: Use both ways. *Correct*: He practiced for long hours and *hardly* (diligently). He *hardly* (seldom) practiced at all. To avoid confusion, sometimes *scarcely* is a better word choice.

have got: Use *have. Correct*: This *has* something wrong with it *(Avoid*: This *has got* something wrong with it).

however: See *but.* Also note that it is generally considered weak to begin a sentence with *However.* Instead, it should follow a semicolon. *Correct*: He ran the mile in under four minutes; *however*, there was no official timer present.

imply: Never confuse with *infer*; they have different meanings. To *imply* means to suggest, whereas to *infer* means to guess or conclude. *Correct*: When he said that, he *implied* that the country was at fault. *Correct*: From the speaker's remarks the lady in the audience was able to *infer* that the speaker thought the President was to blame.

in order that: Requires the subjunctive *may* or *might. Correct*: He gave me the book *in order that I might* improve my understanding of ecology.

infer: See *imply.*

insupportable: Preferable to *unsupportable*, though both are correct.

irregardless: A persisting version of an imaginary and illiterate word. Use *regardless.*

it being: An awkward substitute for *since.*

kind of: Use *somewhat* or *rather.*

lay, lie: These commonly misused words are not difficult to understand. The verb *to lay* means to place, put; the past tense is *laid.* The verb *to lie* means to be at or come to rest; its past forms are *lay* and *have lain* (but not *laid*). *Correct*: We *laid* the books on the table. *Correct*: We *lie* on the blanket in the sunshine. *Correct*: We *lay* on the blanket and *had lain* there on Friday too. Obviously errors can produce results which are either humorous or embarrassing.

lead: The past tense is *led* (not *lead*). *Correct*: We *led* the dog to its master.

leave, let: The verb *to leave* means to depart, go away; the verb *let* means to permit, allow. *Correct*: When are you *leaving* the house to go to school? *Correct*: Will you *let* me *leave* the house now?

leisurely: Both an adverb and an adjective. *Correct*: He strolled *leisurely* through the park. *Correct*: He strolled through the park in a *leisurely* way.

like: If you do want good grammar, and good taste, try to remember that *like* cannot be used as a conjunction. (*Incorrect*: You do it just *like* I do it.) *Correct*: You do it just *as* I do it. Sometimes *like* can be used as an adverb equivalent. *Correct*: You, *like* John, do that all the time. *Like* can also be used as a prepositional adverb. *Correct:* You play *like* a champion.

likely: Used as an adverb, combine with *very, more*, or *most*. *Correct*: It is *very likely* that you will win the contest. Used as an adjective, the word simply means probable. *Correct*: (Sarcastically) That's a *likely* story.

liven, liven up: Use *enliven* instead.

lose, loose: The verb is to *lose* when the meaning is to miss, to be unable to find; the verb is to *loose* when the meaning is to untie, to unbind, or to make *loose*; to *loosen* means to set free.

M.S.: *M.S.* is always capitalized, and this is the singular, the abbreviation for one manuscript; for several manuscripts, use the abbreviation *MSS*.

meantime, meanwhile: The first may be used as a noun as well as an adverb; the second is an adverb only. *Correct*: In the *meantime*, why not do some homework? *Correct*: Run to the store *meanwhile* and see if they have any cookies.

middle class: As a noun, two words. *Correct*: He belongs to the *middle class*. When used as an adjective, a hyphen is required. *Correct*: *middle-class* morality.

naïve, naïveté or naïvty: Spelled this way.

needless to say: This curiously ridiculous phrase is self-defeating. If it is truly *needless* to say something, then why say it at all?

off of: Omit the *of*. *Correct*: He got *off* the bus.

onto: Usually *on* will do the job by itself; however, if *onto* is definitely needed, be sure to spell it as one word.

other, each other: *Each other* in the possessive becomes *each other's*.

otherwise: Try to use only as an adverb (*other* is the adjective).

ought to of: Should be *ought to have*.

our, ours: Do not use *ours* as a preceding compound possessive pronoun. (*Incorrect*: Ours and Russia's space programs are similar.) Shift the *ours* to the end. *Correct*: Russia's space program and *ours* are similar.

over-all: Best hyphenated.

parallel: Spelled as is, can take *to* or *with*.

parentheses: This is the plural of the singular *parenthesis*.

percent: Interchangeable with *per cent* and *per cent.*, but the single word is most commonly used.

phenomena: This is the plural of *phenomenon*.

precipitous: Used primarily to describe physical objects, this adjective means steep, resembling a precipice; to indicate impulsive or rash action, use the adjective *precipitate*.

premise: The plural is *premises*.

principle, principal: *Principal* is both an adjective and a noun, whereas *principle* is a noun only. *Principal*, the adjective, means chief, main; as a noun, *principal* means leader (and, in investment terminology, capital). A *principle* is a law or a truth.

psychotic, neurotic: Do not use these terms interchangeably. *Psychotic* denotes a serious mental imbalance, often bordering on insanity, and thus implies a definite sickness; *neurotic* is the more general word and denotes highly nervous. Many more people are *neurotic* than *psychotic*; the latter term is reserved for the truly sick. *Psychoneurotic* denotes a sickness without an identifiable organic cause. *Psychic*, unrelated to these other words in meaning, denotes clairvoyant or else refers to the psyche. *Psycho* is a colloquialism which attempts to blend all of these words into one meaningless term which gained popularity following Alfred Hitchcock's famous horror movie of that title.

question of: Make sure that there really is a question before using this phrase.

quote, quotation: To *quote* is the verb; *quotation* is the noun. In writing try to avoid the colloquial use of *quote* to stand for *quotation.*

reason: *Avoid*: The reason is because . . . This is redundant because *the reason is* and *because* mean the same thing.

recommend: So spelled.

reinforce: This is the preferred American spelling (not *re-enforce*).

relatively: To use the word *relatively* by itself is silly. If you write, "He ran relatively fast," the reader has no idea how fast he ran—relative to what? the speed of light? his little brother? Again, it is pointless to write, "It's all relative." Limit the use of *relatively* and *relative* to comparative statements. *Correct*: Of the millions of people eligible for the new health benefits, *relatively* few apply for them.

rhythm: All these letters are needed for correct spelling. Also, use *rhythmical* rather than *rhythmic* in almost all cases.

seeing that: Use *since* or *because.*

seems: Do not use to weaken *is*. If you honestly think that American foreign policy *is* in need of substantial revision, it is a cop-out to write that American foreign policy *seems* to be in need of revision.

so-called: *Correct*: The *so-called* now generation. *Correct*: The Romantic poets, *so called* because of their optimistic faith in man's essential goodness, expressed their ideas not only in their poetry but in their political essays as well. (Notice the removal of the hyphen in the second example.)

straight: Both an adverb and an adjective.

subject of (see *concept of*): Often superfluous. (*Avoid*: The *subject of* history is one which has always puzzled men with theoretical minds.) *Write*: History has always puzzled men with theoretical minds.

subtle, subtile: First spelling should always be used. Note other *correct* forms: *subtler, subtlest, subtly.*

succeed: Should take *in* and a gerund. *Correct*: He *succeeded in carrying* out his orders.
supersede: So spelled.
technical: So spelled. Also note *correct* spelling of *technique* and *technically*.
thusly: Illiterate for *thus*.
ultra: Usually no need for a hyphen when prefixed to another word. *Correct*: *Ultraconservative*.
vice versa: So spelled.

7 MECHANICS

Who can deny that at first glance there is something stark about the title of this chapter? "Mechanics," we repeat. "How barren!" We should realize, however, that the word sends us in two directions: We can think of the mechanics of written language, and we can also think of ourselves, writers, as mechanics. In the first instance we are thinking of the craft, and in the second instance of the craftsmen. The word *mechanics* pulls together the writer and his words, and the more persuasive the would-be writer on ecology wants to be, the better he should know his tools. His terms may be different, but as a writer he can increase his effectiveness by learning more about the mechanics of writing which need to be mastered by writers in all fields.

Familiarizing yourself with the mechanics of writing is not really as odious as it sounds. Medical students groan when they memorize the names of bones, but eventually they are glad to know those names. There is not, from any philosophical point of view, anything unpleasant about *technical* improvement of writing. Writing that is technically "clean" is generally the most respected writing. The best writers need the comma. The mechanics of writing are only disappointing in the sense that they, in themselves, do not take us very far. After correctness and clarity, individual style takes over. Everyone can learn to punctuate correctly, but only some writers are *interesting.* Learning some rules of writing puts us all on an equal basis for about ten minutes; after that, personal stylistic characteristics take command. It is still true, however, that we must all learn some rules before we can even compete. Writing is not a cosmetic art. It is internal strength through precision and clarity.

Quite often a paper, on an ecological topic or on any topic, falls just short of being perfect. The ideas are excellent, the writing quality is generally superior, and the paper as a whole is expertly organized; however, there are a few punctuation mistakes. A writer has accidentally placed a sentence fragment that lacks a verb after a semicolon, or he has failed to indent and type single-spaced a lengthy quotation. Often the difference between an A paper and a B-plus-ish one is simply punctuation. Why not shoot for 100 percent accuracy in your punctuation?

Most textbooks divide punctuation into *end* punctuation and *internal* punctuation. Most writers find no difficulty in connection with the former; we have always known how to end different kinds of sentences, and it may be superfluous to discuss the rule for putting a question mark at the end of a question. This still leaves us with internal punctuation—the comma, dash, parentheses, semicolon, and colon.

Use of the Comma

For the most part, the writer needs the comma when he is joining two clauses, both of which have a separate function or may be said to be *independent* clauses. If X and Y are independent clauses, the comma would be used in ways such as the following:

X, and Y.
X, but Y.
X, or y.
X, for (because) Y.
X, yet Y.

All of these coordinating conjunctions must be preceded by a comma when two independent clauses are being joined. A few sentence examples follow:

John could run for the job of Ecology Adviser of Morris County, or he could run for the office of Ecological Consultant of Harville.

John was not eligible to run for mayor, for it had been decided by a right-wing group that only veterans could run.

A writer uses a comma frequently. He usually needs one, for example, when he begins a sentence with a subordinate clause:

When X, Y.
If X, then Y.

When I saw the flag come down, I knew that we had lost.

If John can bring his college-board scores up, he will probably be admitted to Yale.

A short introductory phrase should be set off by a comma:

Granted that, can't we still say that she is smart?

After the slander in the press, Mayor Rebus grew hostile.

Disappointed, she turned her eyes away from him.

When a clause placed later in the sentence qualifies the main clause, it often needs to be preceded by a comma. This is particularly true of a delayed "because" clause:

She knew that he was not serious, because he was rarely serious with any girl.

Some parts of a sentence are restrictive and some nonrestrictive. What is restrictive is the inherent part of the meaning and flow of the sentence; what is nonrestrictive is additional, "tacked on," a qualification which is independent of the structure of the rest of the sentence. Consider the following:

Restrictive: The man who acted as chairman of the committee was not afraid to speak his mind.

Nonrestrictive: John Smith, who acted as chairman of the committee, was not afraid to speak his mind.

In the first sentence the clause stating that the man is chairman is restrictive because it identifies the man, and that person's identity is part of the meaning of the sentence. In the second sentence, the writer is giving additional information about John Smith, which is not an inherent part of the sentence. The clause in this instance, then, is nonrestrictive and is therefore separated by commas. This is one of the thorniest problems of internal punctuation; a few more examples should help to make the difference clear:

Restrictive: A girl who dresses neatly will be asked to few dances.

Nonrestrictive: Barbara Rush, who dresses neatly, will not be asked to many dances.

Restrictive: All of his books written while he was in jail are about the problem of syndicated crime.

Nonrestrictive: His well-known books, which were written when he was in jail, are about the problem of syndicated crime.

Commas are needed when appositives, which are a form of nonrestrictive clause, are included. Consider the following correct examples:

John Wilkes, an aging man, was bitter about the results.

Dickens' *Bleak House*, a contemporary picture of the gloominess of nineteenth-century industrial England, has long been considered a symbolic novel.

Alice, the youngest child there, was very petulant.

Commas are needed to set off any word or expression that is not directly necessary to the rest of the sentence:

Girls, unlike boys, often are forced to go to dancing lessons when they are eleven.

I am telling you for the last time, Harry Wilson, if you won't at least try smoking grass, we're through!

Commas are needed to set off words or phrases in a series:

She liked apples, peaches, cherries, and pears.

There is no difficulty in buying, leasing, selling, or even replacing the equipment.

She is a messy, homely, uncouth girl.

Be careful not to use a final comma between the last of a series of adjectives and the noun they all modify; in the last example you would not want a comma after *uncouth.*

Commas are needed after the conventional transitional words: *however, therefore, moreover, nevertheless, then, consequently, accordingly.* Commas are almost always used when explaining a contrast, as in the example of "girls, unlike boys . . ."

These are the principal times when commas are needed, and most mistakes of comma punctuation arise out of using one when it is not necessary. There are, of course, a few instances in which a comma should be used simply to make the syntax of the sentence more immediately clear. Using a comma for clarity is similar to hyphenating *coworker,* which we discussed earlier. Such instances are rare, and the student will do better to limit his use of the comma to the situations we have described.

Use of the Dash

In general a writer uses a dash to achieve an immediate isolation or suspension of a word or idea, thus making it momentarily special by separating it from the flow of the sentence. When used carefully, the dash is a very convenient tool for emphasizing something. There are, of course, two places in the sentence where the dash may be used: The writer can isolate a word or phrase in the middle of the sentence by setting it between two dashes, or he can isolate a thought or word at the end of the sentence by setting it off with one dash:

The errors—and there are plenty of them—can all be eliminated through careful rewriting.

When you have climbed Mount Everest, you know you have climbed a mountain—and that's the truth!

Sometimes a dash is used effectively when a writer wishes to repeat something with special emphasis:

The whale is a mammal—I repeat, a mammal—that is unconventional in every way.

Perhaps someday—but only someday—the country will be ready for this kind of cultural change.

Usually the material set off by a pair of dashes will explain what immediately precedes it:

Astronomy—which is really the *art* of examining the heavens—is a field with great attraction for young men with sensitive natures.

The three ways of dealing with this problem—execution, delay, and rearrangement—are all valid.

Finally, note that a dash can be used instead of "to" between dates, although here only one dash of the typewriter is used:

In the period 1963–1965, there were great advances in the American economy.

Remember that when you type your manuscript, a dash should be represented by two dash marks; a single dash is, of course, what we use for hyphenated expressions and thus if carelessly used to set off a word or a phrase it confuses the reader.

Use of Parentheses

One of the most interesting aspects of the use of parentheses (and there are not many) is that almost anything included in parentheses could be set off by commas instead. Consider, for example, the last sentence. The phrase "and there are not many" could have been set off either by dashes or commas with little difference. The author in this instance wanted to add a little comment, but not in an overly serious way; so he placed it in parentheses. The decision to place a word or phrase in parentheses usually involves a value judgment, although sometimes parentheses are needed to cite references to page numbers:

> O'Neill makes the point that man always has one more illusion remaining when Hickey suddenly begins to think that he may have been insane when he killed his wife Evelyn (p. 135).

Parentheses are also useful when a writer wishes to translate a few words from a foreign language:

> She shrugged off his criticism with an attitude of *che sera, sera* (what will be, will be).

> The philosophy of *cogito ergo sum* (I think, therefore I am) has always had a strong appeal to young intellectuals.

Whether a writer chooses to use dashes or parentheses to set something off depends on how much attention he wants to call to it; as a general rule, placing something in parentheses calls less attention than does placing it in dashes. Ironically, the least attention is called to something when it is simply set off by commas and included in the sentence.

Use of the Semicolon

The semicolon probably poses almost as much difficulty as the comma. Most students simply never learn the basic rule: what follows a semicolon must read as if it is a complete sentence, and must have, therefore, a subject and a verb:

> *Incorrect*: She was willing to leave school; depending on her parents' reaction to the idea.

> *Correct*: She was willing to leave school; she simply wanted to know what her parents' reaction would be.

In general, the semicolon is used simply to separate the clauses of a compound sentence:

> A penny saved is a penny earned; a penny spent is a penny lost.

Mechanics 99

A semicolon also is used to separate a series of clauses or phrases that have been introduced by a colon:

There have been various times in American history when this was true: 1942–1943; 1956–1958; and 1965–1967.

A semicolon usually precedes certain qualifying expressions or clauses which in some way introduce the opposite of what is said.

There was nothing more beautiful in the room than the painting; however, when the lights went out, nothing could be seen.

She was the kind of woman anyone could have chosen for a wife; that is to say, she was kind, considerate, unselfish, intelligent, beautiful, and affectionate.

Certain key phrases, most of them transitional connectives, are useful to introduce the part of a sentence following a semicolon. These include *namely, that is, for instance, for example, accordingly, also, consequently, furthermore, hence, indeed, moreover, nevertheless, so, still, likewise, otherwise, then, therefore, thus*, and *yet*. All of these words, most of which are transitional connectives, are useful to introduce the part of the sentence following the semicolon. They do not have to be used but they are all useful in organizing the rest of the sentence.

The most basic aspect of the use of the semicolon, then, is that it replaces a conjunction when joining two clauses. It is used frequently for making sharper contrasts between the ideas of separate clauses.

Use of the Colon

The colon presents less difficulty than the comma and the semicolon for most writers, largely because the colon is a more "technical" punctuation mark. Its uses are easier to remember and it is difficult to use it where it does not belong.

The most basic characteristic of the colon is that it *introduces* whatever follows it. It is a direct introduction. Notice, for example, how a writer punctuates his introduction of a quotation:

This can be summarized by a phrase from Milton: "to airy thinness beat."

Jones makes this point in the following statement: "I never was able to submit to government by force."

The introductory use of the colon is most frequently associated with the use of the phrase "the following."

His list included the following chores: setting the table, taking out the garbage, cleaning the garage, mowing the lawn.

As an example of a split infinitive, consider the following: He tried to accidentally let it happen.

A colon is effectively employed to present the reiteration of an idea:

This kind of government does more for the poor: Democracy is designed to benefit the masses.

The colon has certain technical uses such as following the salutation in a letter (Dear Sir:), or between Biblical chapter and verse numbers (Job 12:14), and between hours, minutes, and seconds (2:14). But in general the writer will find himself using the colon mainly to introduce a list or pair of items.

Correct Forms of Quotation

When we consider the mechanics of quotation, we are involved in several problems. First, we should know when to incorporate a quotation directly into our text and when to set it off in single-spaced type. The basic rule is that when *more than two lines are being quoted*, they should be set off in the middle of the page, typed single-spaced, and the quotation marks should be omitted (since it is obvious that the writer is quoting). When a quotation is two lines or less it can be included in the text and, in cases of lines of poetry, the second line separated from the first by the use of a single, slanted dash (/). Consider the following correct examples. In the first, the quotation is incorporated in the text; in the second, the quotation is set off by indentation:

When Barry Commoner wrote, "This is the staggering size of one state's environmental problems," he was speaking of California. Marston Bates, complaining about man's treatment of nature, complained of widespread hunting: "The sport of kings and noblemen has now become the sport of millions, of anyone with an automobile and a rifle or shotgun."

Murray Bookchin summarized the problem well when he wrote:

> Modern man's despoliation of the environment is global in scope, like his imperialisms. It is even extraterrestrial, as witness the disturbances of the Van Allen Belt a few years ago. Human parasitism, today, disrupts not only the atmosphere, climate, water resources, soil, flora, and fauna of a region; it upsets virtually all the basic cycles of nature and threatens to undermine the stability of the environment on a worldwide scale.

PUNCTUATION OF QUOTATIONS

Quotation also involves punctuation. From the previous examples, we can see that a long, indented passage is usually introduced by a colon, whereas a quotation run into your own text should simply be punctuated as if it were your own. Note that in direct quotation, the comma should precede the quotation mark:

She kept saying, "I hate this town."

In general all punctuation marks precede the quotation marks; for example, a closing period is inside (precedes) the closing quotation mark:

She was tired of his tendency to label every opinion "congressional testimony."

She felt that when Browning wrote, "all's right with the world," he was presenting us only with the attitude of a young girl and not with his own view.

When a question makes use of a quotation which itself is not a question, the question mark should follow the final quotation mark:

Is there any other way we can understand Browning's statement that "God's in His heaven"?

If the quotation itself is a question, then, of course, the quotation mark stays outside the question mark:

He asked, "Is there no one willing to do this?"

USE OF QUOTATION MARKS

The final problem we have here concerns *when* to use quotation marks. First, you might remember at the outset that book titles are never put in quotation marks; they are always underlined (italicized), as are titles of newspapers and major plays. Short plays, poems, articles, and parts of books are all placed in quotation marks:

Have you read Robert Frost's poem "The Oven Bird"?

Have you read Kenneth Boulding's article entitled "The Economics of the Coming Spaceship Earth"? It appeared in a book edited by Henry Jarrett entitled *Environmental Quality in a Growing Economy.*

All direct quotations are, of course, placed in quotation marks. For quotations within quotations, single quotation marks are used:

She said, "What John said—and I remember his exact words—was, 'I don't want to go,' and not 'I won't go.' "

Sometimes quotation marks can be used to set off a word which is unfamiliar or technical:

When we talk about the earth's "ecosystem," we are talking about the whole complex web of biological processes that make up the universe.

The writer should not extend this use of quotation marks to the introduction of cute expressions and clichés; it is incorrect, for example, to use quotation marks in this way:

His own personal "hang-up" was that he could not "stand" his brother.

It is correct to use quotation marks when referring to a word *as a word*:

The word "love" has so many different meanings that it is impossible to isolate any one of them as a universal definition. What differentiates "love" from "like" is not always clear; many people seem to use the two words almost interchangeably.

(It is also acceptable to italicize words used as words.)

Finally, when a word has a special meaning it sometimes needs to be kept in quotation marks:

There was a "wildcat" strike at the car factory.

"Yellow journalism" is supposedly dead in the United States.

This use of quotation marks should be restricted to terms that are definitely of a special nature on a universal basis.

A few words about the art of quotation are needed. Remember that no good writer quotes too much at a time. A writer simply cannot have one paragraph written by someone else for every paragraph of his own. Moreover, too many short quotations within a paragraph become cumbersome and cut into the writer's effectiveness. Finally, remember that it is sometimes better to summarize another writer's attitude, idea, or position (giving credit) than it is to quote from his writings in a piecemeal fashion. Sometimes authors will not allow things to be quoted from their books, because the statements, when taken out of their larger context, may be misconstrued.

Capitalization

The rules of capitalization are lengthy and important to the writer who wants to be precise and at least semiprofessional. In general, any word with specific rather than general reference must be capitalized. We need to be able to differentiate immediately between a river and the Amazon River. Before listing the words that should be capitalized, we note that when a derivative of a proper noun is used in a secondary sense, it does not need to be capitalized. We write "Utopia" but "utopian ideas."

The following should be capitalized:

1. Names of peoples (French).
2. Names of races (Caucasian).
3. Names of tribes (Navajo).
4. Names of languages (English).
5. Titles of honor when accompanying proper nouns (King James).
6. Academic titles when accompanying proper nouns (Dean Smith).
7. Religious titles (the Reverend John Smith, the Reverend Mr. Smith, the Reverend Dr. Smith).
8. Business titles when accompanying proper nouns (John Smith, Treasurer).
9. Titles of government officials when accompanying proper nouns (Senator Hatfield).
10. Names of national or international bodies and documents of the government (the Security Council of the United Nations, the Constitution of the United States, the Warsaw Pact).

11. Nouns and pronouns referring to the deity (God, Him).
12. Names for the Bible and its parts (Job, the Scriptures, the Bible).
13. Names of religious orders, creeds, and confessions (Carmelites, the Apostles' Creed).
14. The months, days, holidays, holy days (October, Tuesday, Thanksgiving, Easter).
15. Names of geological eras, periods, epochs (Carboniferous, Age of Reptiles, Neolithic Age).
16. Names of class names but not of species (the Coelenterata, but a coelenterate).
17. Names of planets, constellations, asteroids, stars, and groups of stars (Mars, Orion's Belt, the Big Dipper, the Milky Way), but not sun, moon, or earth unless in connection with other astronomical terms ("The sun is shining today," but "How does the Sun relate to Jupiter?").
18. Generic geographical terms (Atlantic Ocean, the Southern Hemisphere).
19. Political parties and political generic terms (the Republicans and the Independents, the Holy Roman Empire).
20. The points of the compass when used to designate particular political entities but not when simply used to indicate direction (Southeast Asia, but, "let's drive due north.")
21. All the words in titles of books, periodicals, essays, and poems except for prepositions and articles (*The Turn of the Screw, Moby Dick, "Tintern Abbey"*). "The" may be capitalized when it precedes the name of a newspaper if it is part of the newspaper's name (the *Boston Globe*, but *The New York Times*).

Finally, remember that it is weak to capitalize simply for emphasis. As should be obvious by now, words are capitalized for particular reasons according to an established set of rules which can be found in the back of most college dictionaries. A writer cannot, therefore, capitalize a word just for effect, cannot write, "Don't lose your Cool," in order to make the reader gasp out the word "cool" in a more exhilarating way.

Rules for Spelling

Many students labor under the false assumption that correct spelling is only a matter of looking up words in a dictionary. There is no denying that the best way of becoming a good speller *is* by looking up words in a dictionary. But there are also basic linguistic rules that enable the student to spell *logically* when there is no dictionary available.

1. There is a conventional rhyme which many students learn in grade school and which all students should relearn in college: "Use *i* before *e* except after *c* or when used as an *a* as in *neighbors* and *weigh*." Examples: receive, deceive, believe, weight. Exception: leisure.
2. Plurals of nouns ending in *i* are usually formed simply by adding *s*. Examples: rabbis, alibis.
3. Nouns ending in *y* preceded by a consonant and those ending in *y* preceded by *u* pronounced as a consonant are pluralized by changing the *y* to *i* and adding *es*. Examples: sky, skies; fly, flies; mercy, mercies; colloquy, colloquies. Notice, however, that this rule does *not* pertain to proper names. We write: There are three Bettys and two Bobbys in the class.
4. Nouns ending in *y* preceded by *a, e, o,* or *u* are pluralized simply by adding an *s*. Examples: way, ways; key, keys; buy, buys.
5. Words ending in a double consonant generally retain both letters when pluralized or lengthened. Examples: will, wills; willing (never wiling, for example); jazz, jazzy.
6. Words ending with a silent *e* usually retain the *e* when a suffix beginning with a consonant is added and drop the *e* when the suffix begins with a vowel. Examples: come, coming; drive, driving; live, living; pale, paleness. There are exceptions to this general rule, as, for example, with various -*able* words such as *noticeable*.
7. Words ending in *c* have a *k* inserted before the word is lengthened by suffixes beginning with *e, i,* or *y*. Example: panic, panicked, panicky; picnic, picnicked.
8. Usually a word ending in *y* preceded by a consonant will change the *y* to *i* before the word is lengthened in any way. Examples: delinquency, delinquencies; mercy, merciless.
9. Usually a word ending in *y* preceded by a vowel will *not* change the *y* to an *i* when the word is lengthened. Examples: obey, obeying; play, player.
10. A final rule to remember is that when a word is formed (compound word) out of other words, the general rule is that all of the letters of both words are retained. Examples: split-level, airstrip, fountainhead.

There are some words that everyone seems to misspell often. To compile a list of all the words that we misspell would, of course, result in reproducing the dictionary. The following list of *correctly* spelled words has been assembled somewhat arbitrarily, but it is hoped that the most commonly misspelled words are here:

accept	already
accommodate	apparent
achieve	believe
all ready	category
all right	choose

climactic	repetition
conscience	rhythm
decent	sense
dual	separate
existence	sight (site, cite)
explanation	similar
its (it's)	stationary (stationery)
height	succeed
loose	success
losing	technical
necessary	technique
occurred	than (then)
occurrence	their (they're)
occurring	thorough
of, off	to (too, two)
peace (piece)	unnecessary
precede	woman (plural=women)
privilege	writing
professor	your (you're)
recommend	

As can be seen, there are many times when words are misspelled sheerly by accident. A writer accidentally spells "to" "too." Such phonetic spelling errors are common even in professionally prepared manuscripts. Proofreading is the only solution. It is also interesting to note that one of the most frequently misspelled words is "misspelled."

Use of the Hyphen

When two words are serving as a combined modifier and they precede the term modified, they are joined by a hyphen. When the two words follow the word modified, they are kept separate:

Paul Ehrlich is *well known* for his writings on the environment.

The *well-known* environmentalist Paul Ehrlich gave an exciting speech.

The *well-dressed* speaker stood up.

When the speaker stood up it was clear that he was *well dressed.*

Use of the Apostrophe

When a word does not end in *s*, the possessive is formed by adding an apostrophe and an *s*, as when writing a *woman's* book, a *boy's* hat. When a singular word ends in *s*, it is usually correct to add an apostrophe and an *s* to form the possessive.

It is correct to write *Chris's* book. Sometimes this will appear awkward or make the word difficult to pronounce, in which case one simply uses the apostrophe without the additional *s*, as in writing *Socrates'* death. When the plural of a word ends in an *s*, however, it is *never* correct to add an *s*; instead, use the apostrophe by itself, as when writing the *Hebrides'* woolen goods, or the *boys'* shirts. Remember, too, that an apostrophe and an *s* should be used to pluralize letters, numbers, and words when the words are discussed as words. Thus we correctly write:

> Please eliminate the *3's* and the *5's*. The *6's* may stay. Be sure to go over your paper to see if all of your *i's* have been dotted.

> There are too many *if's*, *and's*, and *but's* here.

Italics

When writing a paper, indicate italics by underlining (remember, do not use quotation marks). Titles of books, newspapers, magazines, plays, and long poems should be italicized:

> *Paradise Lost*
> *Time*
> *The Territorial Imperative*
> *Newsweek*
> the *Saturday Evening Post*
> the *Boston Globe*
> *Our Town*

Also remember to italicize works of art and movies—Rembrandt's *Aristotle Contemplating the Bust of Homer*, *Easy Rider*, *The Andromeda Strain*. Although it is generally preferred to place words used as words in quotation marks, it is also acceptable to italicize them. Numbers and letters used as numbers and letters should be italicized:

> There are too many *and's* in this paragraph.

How many *r's* are there in *correspondence*?

Sometimes your *8's* are impossible to recognize.

Finally, italicize foreign words:

The point of this article is summarized by the Italian motto *che sera, sera* (what will be, will be).

BIBLIOGRAPHY

Bates, Marston. *The Forest and the Sea.* New York: Random House, 1960.
Blau, Sheridan, and Rodenbeck, John. *The House We Live In.* New York: Macmillan, 1971.
Bookchin, Murray. "Ecology and Revolutionary Thought," *Anarchos,* Vol. 1 (February 1968).
Boulding, Kenneth E. "The Economics of the Coming Spaceship Earth," in Henry Jarrett (ed.), *Environmental Quality in a Growing Economy.* Baltimore: Johns Hopkins Press, 1966.
Bowner, Ernest J. "Salmonella in Food: A Review," *Journal of Milk and Food Technology,* Vol. 28 (1965).
Bruce, F. E. "Water Supply, Sanitation and Dispersal of Waste Matter," in W. Hobson (ed.), *Theory and Practice of Public Health.* London: Oxford University Press, 1961.
Carson, Rachel. *The Edge of the Sea.* Boston: Houghton Mifflin, 1955.
―――. *Silent Spring.* Boston: Houghton Mifflin, 1962.
Chute, Robert M. *Environmental Insight.* New York: Harper & Row, 1971.
Commoner, Barry. *Science and Survival.* New York: The Viking Press, 1966.
―――. "Soil and Fresh Water: Damaged Global Fabric," *Environment,* Vol. 12, No. 3 (April 1970).
Curtis, Richard, and Hogan, Elizabeth. "The Myth of the Peaceful Atom," *Natural History* (March 1969).
Dubos, Rene. *The Mirage of Health.* New York: Doubleday Anchor, 1961.
―――. *Reason Awake.* New York: Columbia University Press, 1971.
Eastman Kodak Co. *Improve Your Environment . . . Fight Pollution with Pictures.* A. C. 26. Rochester, New York, 1971.
Ehrlich, Paul R. *The Population Bomb.* New York: Ballantine Books, 1968.
First, Melvin W., Viles, F. J., and Levin, S., "Control of Toxic and Explosive Hazards in Buildings Erected on Landfills," *Public Health Reports,* Vol. 81 (1966).
Fischer, John. "Survival U: Prospectus for a Really Relevant University," *Harpers Magazine* (September 1969).
Food and Agriculture Organization of the United Nations. *Evaluation of the Toxicity of Pesticide Residues in Foods.* Rome, 1965.
Fuller, R. Buckminster. *Ideas and Integrities.* Englewood Cliffs, N.J.: Prentice-Hall, 1963.
Graham, Frank, Jr. *Since Silent Spring.* Boston: Houghton Mifflin, 1970.
―――. Man's Dominion, M. Evans, New York, 1971.
Grossman, Shelly. *Understanding Ecology.* New York: Grosset and Dunlap, 1970.
Haagen-Smit, Arie J. "The Troubled Outdoors," in Burgess Hill Jennings and

John Edward Murphy. *Interactions of Man and His Environment.* New York: Plenum Publishing Corp., 1966.

Hanley, Amos. *Human Ecology: A Theory of Community Structure.* New York: Ronald Press, 1950.

Hanson, Herbert C. *Dictionary of Ecology.* Washington: Catholic University of America Philosophical Library, 1960.

Hare, Kenneth F. "How Should We Treat Environment?", *Science,* Vol. 167 (Jan. 23, 1970).

Harte, John, and Socolow, Robert H. *Patient Earth.* New York: Holt, Rinehart and Winston, 1971.

Helfrich, Harold W., Jr. (ed.) *Agenda for Survival.* New Haven: Yale University Press, 1970.

Jarrett, Henry (ed.) *Environmental Quality in a Growing Economy.* Baltimore: Johns Hopkins Press, 1966.

Love, Rhoda M., and Love, Glen A. *Ecological Crisis.* New York: Harcourt Brace, 1970.

Marx, Wesley. *The Frail Ocean.* New York: Coward McCann, 1967.

Means, Richard L. *The Ethical Imperative.* New York: Doubleday, 1969.

Mitchell, John G., and Stallings, Constance L. (eds.). *Ecotactics.* New York: Pocket Books, 1970.

Moore, John. *Science for Society—A Bibliography.* Washington, D.C.: American Association for the Advancement of Science, 1971.

Murdoch, William, and Connell, Joseph. "All About Ecology," *Center Magazine* (January 1970) (a publication of the center for the study of Democratic Institutions in Santa Barbara).

Nader, Ralph. "Man-Made Environmental Hazards," *N.J.E.A. Review* (December 1970), 180 W. State Street, Trenton, N.J. 08608.

Novick, Sheldon. *The Careless Atom.* Boston: Houghton Mifflin, 1969.

Olds, Jerome. "Earth Day, How to Make It Survive," *Organic Gardening* (April 1970).

President's Science Advisory Committee, Environmental Pollution Panel. *Restoring the Quality of Our Environment.* Washington, 1965.

Porter, Van Rensselaer. *Bioethics: Bridge to the Future.* Englewood Cliffs, New Jersey: Prentice-Hall, 1971.

Shepard, Paul, and McKinley, Daniel. *The Subversive Science.* Boston: Houghton Mifflin, 1969.

Strobbe, Maurice. *Understanding Environmental Pollution.* St. Louis, Mo.: C. V. Mosby, 1971.

Top, Franklin H. "Environment in Relation to Infectious Diseases," *Archives of Environmental Health,* Vol. 9 (1964).

Udall, Stewart. *The Quiet Crisis.* New York: Holt, Rinehart and Winston, 1963.

Wager, J. Alan. "The Challenge of Environmental Education," *Today's Education* (December, 1970).

Ways, Max. *The Environment: A National Mission for the Seventies.* New York: Harper & Row, 1970.

Westman, James R. "The Imbalance of Nature," *New York State Conservationist* (December-January, 1966-1967).

Wheeler, Harvey. "The Politics of Ecology," *Saturday Review* (March 7, 1970).
White, Lynn, Jr. "The Historical Roots of Our Ecological Crisis," *Science,* Vol. 155 (March 10, 1967).
Wise, William. *Killer Smog.* Chicago: Rand McNally, 1968.

ABOUT THE AUTHORS

Mary Travis Arny is Associate Professor of Biology at Montclair State College. She received the M.Sc. degree from Rutgers University and has been studying ecology and wildlife since the 1920s. Her travels have taken her through the United States, Canada, Bermuda, Britain, Mexico, France, and most recently Africa. Her published works include *Red Lion Rampant, Gardening Guide Book, A Goodly Heritage, Seasoned With Salt,* and *Birds and Mammals of Montclair.*

Christopher R. Reaske, Assistant Professor of English Language and Literature and Director of Freshman English at the University of Michigan, received the Ph.D. and A.M. degrees from Harvard. He is the author of *The College Writer's Guide to the Study of Literature*, the editor of *Seven Essayists*, co-editor of *Student Voices / One* and of the forthcoming *Mirrors: An Introduction to Literature*, and has been a contributor to the *Saturday Review, Shakespeare Quarterly, Journal of the History of Ideas, Michigan Academician*, and *Anon.* Christopher Reaske is Mary Arny's son-in-law.

This book is printed on Earthtone Offset,
100% recycled paper.

DATE DUE

#47-0108 Peel Off Pressure Sensitive